T0179259

COMPUTER INTENSIVE STATISTICAL METHODS

VALIDATION
MODEL SELECTION
AND BOOTSTRAP

J.S. Urban Hjorth

Department of Mathematics
Linköping University
Sweden

CRC Press
Taylor & Francis Group
Boca Raton London New York

CRC Press is an imprint of the
Taylor & Francis Group, an **informa** business
A CHAPMAN & HALL BOOK

Library of Congress Cataloging-in-Publication Data

Hjorth, J. S. Urban.
 Computer intensive statistical methods : validation model selection
and bootstrap / J.S. Urban Hjorth.
  p. cm.
 Originally published: London ; New York : Chapman & Hall, 1994.
 Includes bibliographical references and index.
 ISBN 0-412-49160-5 (alk. paper)
 1. Mathematical statistics—Data processing. I. title.
QA276.4.H56 1999
519.5'.0285—dc21   99-29714

Originally published by Chapman & Hall
First edition 1994

CRC Press
Taylor & Francis Group
6000 Broken Sound Parkway NW, Suite 300
Boca Raton, FL 33487-2742

ISBN-13: 978-0-412-49160-3 (hbk)

Contents

Preface

The advance of computation has put statistics in a new perspective. Idealized model assumptions can now be replaced by more realistic modelling or by more or less model free analyses. Much statistical work and data analysis is now made by computers in ways that are too complicated for realistic analytical treatment. Automatic model selection poses for example new statistical questions which have been around for some time, but which recently have found working solutions. The statistical properties of the results from these and other extensive computations may well be different from the results from classical analyses. These problems are present in most regression and time series modelling and also in classification, clustering and the like. The new effects caused by all this computation can be approached by a further round of computations as we do in validation and bootstrap methods. Probably this is the only general way to proceed, but the methods will of course continue to be developed.

The enormous success of bootstrap methodology shows that many tricky problems have awaited something different and have now found a useful approach. Another good reason for the interest is that these methods are both amusing and theoretically appealing.

Classically educated statisticians should take a close look at these methods, and the young generation will certainly be handicapped without them.

This work benefits from theoretical work by many persons. Hopefully I have covered the most important contributions by the given references but I have not attempted to give a complete list of the work in the field. Instead I have referred to the work that fits into the line of this text, and have tried to keep methods and descriptions as easy as possible, as long as possible, to make the text readable for the widest possible audience. This is for example the reason why elementary confidence interval methods are presented in Chapter 5 and come back together with more advanced methods in Chapter 6, where more difficult material is collected.

I want to thank some colleagues and research students for their contributions. Ola Junghard, Anders Nordgaard, Elisabet Schmeling, Charlotte Karlsson, Lars Holmqvist, and Lars Häggmark SMHI have all applied or developed methods used in this text.

The Royal Statistical Society permitted the reproduction of material from Applied Statistics (Hjorth and Holmqvist, 1981), Professor Bradley Efron let me use one of his tables, not to mention all the ideas I took without permission, and Professor Sture Holm allowed me to use material from a regression bootstrap paper to appear. I hope nobody else will find their central ideas here without proper references, but sometimes the same ideas pop up independently at about the same time.

A special thanks is due to the Swedish Transport Research Board for their interest in theoretical development for applications in traffic science and for their financial support in the important last stage of this work.

Urban Hjorth
Linköping

1993

CHAPTER 1

Prelude

1.1 Background

Some powerful and very fundamental new principles have been developed in statistics for model selection, estimation, and evaluation of uncertainty. The underlying ideas are often simple and allow great freedom regarding the model structure. Instead they require repeated, repeated and repeated ... almost identical computations on the data and are therefore often named computer intensive statistical methods.

Being computer intensive is a relative quality. We will here use it relative to traditional statistical computation. This means that the computations are in many cases still small compared to everyday computations in such areas as meteorology, physics or mechanics of materials. There are in fact several small examples where the intensive computations are done after a few seconds or minutes on a small PC. Other problems may require more heavy computations and in some cases the fastest computers will still set the limits of what we can do.

The large interest in computer intensive methods in statistics is basically not founded on the computer as such, nor on computer science in a broader sense. Instead the interest comes from the information concept. The purpose of every classical or new statistical analysis is to take care of and present the information in the given data as efficiently as possible. The means to attain this goal can vary, but will typically require computations which are heavy to do by hand. This motivated an early interest in automatic tools for the computations but the progress was fairly slow. Time and cost considerations therefore directed statistical work towards models and methods which kept down the amount of computation, and it is not until recently that these limitations have been removed almost entirely, and computer intensive statistical methods have spread over the world. Even extensive computations will today give costs which can typically be neglected compared to data collection, theoretical work, report writing and the consequences of the studies.

A lot of computation does, however, not guarantee that the information has been well used. The inference problem must be well understood and a small but smart analysis can be more correct than one which studies 'everything'.

Much statistical software is now available and easy to use without much theoretical feeling for the subject. The potential misuse of statistical measures is therefore larger than ever. This refers in particular to the classical methods, which so far dominate the market, and which are typically based on special parametric assumptions and correct data collection. However, the new computer intensive methods are not infallible in this respect. They may be based on less parametric assumptions, but they certainly need to be based on correct data and they give often more rough and approximate solutions in situations where the classical parametric assumptions are fulfilled. It is therefore important to be critical and have a theoretical view in order to understand when computer intensive methods have anything essential to offer compared to less computer intensive techniques. The methods are justified when they can extract the information and describe the uncertainty better than alternative techniques.

Some frequent misuses of classical statistical software can in fact be counteracted by computer intensive methods, and we will in particular discuss these possibilities in model selection situations later. This means that some problems, in a sense caused by the computer, can also be solved by the computer. At the same time the new methods solve many problems which could not be properly attacked before.

1.2 About models

We will make extensive but abstract use of different statistical models, and for those readers who do not swim among such models all the time the following introduction may be useful.

A very general statement is that we need models in order to structure our ideas and conclusions. The models we use are simplified pictures of the phenomenon we are studying. Except for some very pure situations the models are therefore not 'true' but will typically have some defects.

The computer intensive methods make fewer assumptions than classical methods, and are robust against small departures from these assumptions. One potential use of them is therefore to correct for some model defects such as when estimates are based on oversimplified models. (Many engineering results in reliability and queueing are for example based on assumed exponential distributions, but are much more widely used). But this should not be stressed too hard. Some results presume that the corrections are small compared to the variability (for example bootstrap estimation of variability and bias correction). Sometimes we do not even need an explicit model, although we always feel there is one in the background. A prediction formula or an estimator can be all we need. Conclusions drawn about the models will be transferred to the real phenomenon that we are modelling, and with good models the model defects will not have serious effects on these conclusions.

We will mostly limit ourselves to stochastic models with a set of adjustable constants, parameters, which are unknown or only partially known. When distribution free statistical methods are used, we may even consider more vague model structures which can sometimes be characterized by an infinite dimensional parameter vector.

Statistical models can be regarded as tools for collecting and combining information. Already when a model is formulated we can (and to some extent must) bring in prior knowledge based on experience of the studied object. Ob-

servations and estimates of model parameters will sharpen
this knowledge. One of the advantages of statistical models
is that this sharpening can be described and measured, for
example by the width of confidence intervals for the parameters. But information can be used for more than confidence
intervals. In most models we can also find methods to predict
future results. The validation methods in Chapters 3 and 4
will be based on such predictions and measures of efficiency.
We will strongly emphasize that a parameter estimate or a
forecast is typically not of much value unless we know its
uncertainty, even if mathematics is involved somewhere.

When we have uncertain prior knowledge about the parameters in a model we can possibly state this knowledge
as a probability distribution for the parameters. There is
an extensive literature about Bayesian methods to balance
such information against information from new data. We
will not go into such methods here. If we have access to
observations on which the old information is based, we can
instead make a total model for all the data and estimate
the parameters in this model. This will give an alternative
balancing of all the available information. Bayesian methods are much debated especially when they combine purely
subjective judgements, e.g. by a decision maker, with real
data. The difficulties of translating prior knowledge into distributions for the parameters, the lack of objectivity for the
results (different conclusions from the same data due to the
opinions of decision makers) and difficulties with the inter-
pretations in general have strongly limited the practical use
of Bayesian methods, but they nevertheless have a considerable theoretical interest and some good practical arguments
in recursive situations.

The inexperienced observer may have difficulties in setting up a model for a somewhat complex data situation. For
the more experienced, the main difficulty can be that he
or she sees so many approximately equivalent alternatives.
One must decide upon which part of the model is random
and which is fixed, how to define parameters, which assumptions about independence are reasonable and an appropri-

ate mathematical structure. Certain models have, however, developed into what we may call standard models. These models are supported by an extensive literature which treats their theoretical properties (when they are true) and also by extensive computer software. The reader should, however, not feel restricted to such models. Instead we use the freedom of computer intensive methods as a key to modelling in a way which can be more suitable for the situation under study. We will as preparation discuss different models of the regression type. Later in the text we will also discuss some time series models and models for independent equally distributed variables.

Before we start off, let us agree upon some general notation.

1.2.1 Notation

Where possible, without too much need for notational versions, we will use lower case letters $x, y, z \ldots$ for observed values, not necessarily stochastic, and the corresponding upper case letters $X, Y, Z \ldots$ for their stochastic model versions. The letter ε denotes random error unless otherwise explicitly stated.

The Greek letters $\alpha, \beta, \ldots, \theta \ldots$ and sometimes also the letters $a, b, c \ldots$ denote parameters. θ is preferred as a general patameter, β is preferred for regression coefficients. Estimated values of parameters are denoted by hats like $\hat{\theta}$.

Resampled data and estimates based on the resamples (Chapter 5 and later) are denoted by stars like $x^*, \hat{\theta}^*$, and parameters in the resample distribution are denoted by tilde, like $\tilde{\theta}$.

The upper case letters $F(.), G(.), H(.)$ usually denote distribution functions, and $f(.), g(.), h(.)$ are the corresponding density functions or discrete frequency functions.

For a random variable, say Y, with distribution function $F(y)$, we write $\mu = E[Y] = \int y \, dF(y)$ for the expected value, where as usual the integral is evaluated as a sum if F is discrete and as $\int y f(y) dy$ if F admits a density $f(y)$. We also have the variance $\sigma^2 = E[(Y - \mu)^2] = \int (y - \mu)^2 dF(y)$, and the standard deviation σ.

For a sample $\mathbf{y} = (y_1, \ldots, y_n)$, modelled as independent and identically distributed variables $\mathbf{Y} = (Y_1, \ldots, Y_n)$, we typically estimate μ by $\bar{y} = (y_1 + \ldots + y_n)/n$ and σ by $s = s(\mathbf{y}) = \sqrt{\frac{1}{n-1}\Sigma(y_i - \bar{y})^2}$. The same notation can be used both for the real values computed from the data and for the abstract model version of the data. The context will typically tell which version is appropriate. If we want to emphasize the stochastic version, we can instead use the notation $\bar{Y}$, $s(\mathbf{Y})$.

We write $N(\mu, \sigma)$ as a name for the normal distribution with the density

$$f(x) = \frac{1}{\sqrt{2\pi\sigma^2}}e^{-\frac{1}{2}(\frac{x-\mu}{\sigma})^2},$$

and $\Phi(x)$ for the $N(0, 1)$ distribution function. Also $t(r)$, and $\chi^2(r)$ are names of the Student's t-distribution and the Chi-squared distribution, each with r degrees of freedom.

Other symbols are $\mathrm{Exp}(\mu)$ for the exponential distribution with expected value μ and probability density $f(x) = \frac{1}{\mu}e^{-x/\mu}$, $x > 0$. This distribution is equally often parametrized by its intensity $\lambda = 1/\mu$, so there is a risk of confusion. $\mathrm{Re}(a, b)$ stands for the rectangular (uniform) distribution in the interval (a, b), $a < b$, on the real line, and $\mathrm{Bi}(n, p)$ for the binomial number of successes in n independent attempts with probability p each.

Since these notations are standard in much statistical litterature they are hopefully easy to remember. Further notation and variants of these symbols will be defined when the need arises.

1.2.2 Some model examples

A few models of the regression type will be given here as a background to the model discussions in the following chapters. The main message is to demonstrate the many variations such models can have. As a consequence the need for proper model selection and evaluation methods will be emphasized.

Example 1.1   Standard linear model
A linear model connects a quantity Y to other quantities $x_1, \ldots, x_m$.

$$Y = \beta_0 + x_1\beta_1 + \ldots + x_m\beta_m + \varepsilon, \qquad (1.1)$$

where ε is a random variable, independent of $x_1, \ldots, x_m$ and also independent for different observations of $(Y, x_1, \ldots, x_m)$. The error ε may have a known distribution up to some parameters, and is usually assumed to be normal with mean zero and variance σ^2.

This standard model gives a structure (here linear). Prior knowledge is entered in many ways. The most important is the selection of variables $x_1, \ldots, x_m$ which may be based on physical, economical or other knowledge saying that these quantities ought to be related to Y. Prior knowledge can also be entered through assumptions about ε, through the linear structure, or as side conditions on the parameters. From independent observations on $(Y, x_1, \ldots, x_m)$ the parameters β_i and σ are estimated by regression analysis but we will not enter into such details here.

The strength of the assumptions we have imposed by the linear model is perhaps best seen by discussing some close alternatives. We may have reason to suspect that a combination of large (or small) values in the variables x_1 and x_2 will have a special effect. Probably the simplest adjustment will then be to introduce one or a few new variables e.g. $x_{m+1} = x_1 x_2$ and return to the linear model structure. This means that we are linearizing the non-linear structure at least approximately.

Even if the variables can be treated one at a time, there is often no logical reason why they should enter linearly in the model. If so, we may introduce transforms of the variables such as $g_i(x_i) = x_i^{p_i}$ where p_i can be an arbitrary parameter if x_i is always positive. We then have the following model.

Example 1.2   Model with transformed predictors

$$Y = \beta_0 + \sum_1^m \beta_i x_i^{p_i} + \varepsilon. \qquad (1.2)$$

With the same assumptions on ε as before, this model has $2m + 2$ parameters $\beta_0, \beta_i, p_i, \sigma$ to be estimated.

Many other transforms can be considered. If the x_i have both signs, the transforms $h(x_i) = x_i |x_i|^{p_i - 1}$ are useful substitutes for $x_i^{p_i}$. Such power transforms will give the value $x_i = 0$ a special role, since the effect of a change in x_i will depend on x_i's distance to zero. At the cost of some more parameters (x_{i0}) any value could take this special role. We just replace x_i by $x_i - x_{i0}$ in g_i or h_i. The information about p_i and x_{i0} can be weak. Sometimes it is therefore better to offer a few distinct possible values for these parameters. For polynomials the centring problem may disappear. If for example x_i, x_i^2, x_i^3 are all in the model, no parameter x_{i0} will be needed and yet no particular value is singled out since all expressions $(x_i - x_{i0})^k$, $k = 1, 2, 3$, can be described by proper values of the β-coefficients.

If only changes of x_i in a certain region are important and we have some sort of saturation effects outside this region, a continuous distribution function $F(x)$ can be used to define transforms $l_i(x_i) = F((x_i - \mu_i)/\sigma_i)$. Useful functions are $F(x) = e^x/(1+e^x)$, or the normal distribution function $\Phi(x)$. The values μ_i and σ_i will define the region of interest and become parameters if we do not know this region beforehand.

We can also transform the predictand Y, and if Y is always positive one possibility is to take logarithms $\ln Y$ of Y. Suppose that all variables are positive by definition and that we use their logarithms. We can then arrive at the following model.

Example 1.3   Multiplicative model

$$\ln Y = \ln \beta_0 + \sum \beta_i \ln x_i + \varepsilon \qquad (1.3)$$

or, rewritten,

$$Y = \beta_0 \cdot x_1^{\beta_1} \cdots x_m^{\beta_m} \cdot e^{\varepsilon}.$$

Again we may use the normal distribution for ε if appropriate, or any other distribution with mean zero.

Anyone with experience of linear and non-linear models, or optimization in general, will know how very much simpler it is to estimate the parameters of the linear model (1.1) than those of the non-linear model (1.2). Since the multiplicative model (1.3) is linear in the parameters β_i (set $\beta_0^1 = \ln \beta_0$) we can estimate its parameters as easily as in the model (1.1) if the assumptions about ε are appropriate.

A different generalization, which can often be motivated by a priori knowledge, is that the random deviation ε has a dispersion which depends on the predictors x_i. Such models are usually called heteroscedastic.

Example 1.4   Heteroscedastic model
Assume a model like (1.1), (1.2), or (1.3) for the undisturbed, or average, relation between Y and $x_1, \ldots, x_m$. Let ε be $N(0, \sigma(x_1, \ldots, x_m))$ where $\sigma = \sigma_0 e^{\Sigma \gamma_i x_i}$ or $\sigma = \sigma_0(1 + (\Sigma \gamma_i x_i)^2)$ or some other positive function of the x_i.

One cause for this model is when a Poisson distribution of the variable Y is approximated by the normal distribution (a valid approximation if the expected value is large). Since the Poisson distribution has the same mean and standard deviation, a model for the mean should also be a model for the standard deviation in the normal approximation. Another application was found by the author in meteorology, where situations with different stability caused variation in the uncertainty of temperature forecasts. This time different sets of predictors were used for the mean and for the heteroscedastic part.

One more generalization of the linear model can be of interest here.

Example 1.5   Binary regression
Let Y take the values 0 or 1 and assume that the probability for $Y = 1$ depends on predictors $x_1, \ldots, x_m$ in the following way. Let $G(x)$ be a given monotone function taking values between 0 and 1. (G is typically a distribution function but is used with a different interpretation.) Put

$$P(Y = 1 | x_1, \ldots, x_m) = G(\beta_0 + \sum_1^m \beta_i x_i). \qquad (1.4)$$

This is a generalized linear model for binary Y-variables. Such models have a large theory on their own and are extensively studied in McCullagh and Nelder (1989). In the special case $G(x) = \Phi(x)$ (the normal distribution function) we call (1.4) a probit model, and if $G(x) = e^x/(1 + e^x)$ it is called a logit model.

These five examples from the theory of regression analysis can illustrate the variety of models. Within each example we can also vary the number of predictors x_i and their definitions. Ideally our experience should dictate our model choice before the data are analysed, but in practice we will more often get into situations where the prior information is too vague and many models have to be tried on the same set of observations. Each of these models express one possible opinion of which factors will explain each other and a structure for the relationship. For every such model studied, we will estimate the parameters and also estimate the precision of the estimates and the fit or the predictability of the model. Except for the purely linear model this will typically require numerical solution of some estimation equations for example by maximizing the probability, the likelihood, of the data observed. Mostly the computation of the estimates will be an easy to moderately difficult task. Estimating the precision of the parameter estimates within the framework of a given model is a more difficult problem and in situations where we think the model may be wrong or where we may select between many alternative models the problem becomes troublesome. This class of problems will be our theme in the next chapters.

CHAPTER 2

Computer intensive philosophy

Estimates of a parameter from different data bases will typically give different results. A classical statistical analysis of one such estimate on a given data base starts from a model giving a probability measure for the set of all conceivable states of this data base. From that model a probability distribution is derived theoretically for the variations of the estimate. This allows conclusions about how far from the true parameter our estimate may fall with a certain probability. On that basis statements are made about the true parameter.

The new contribution to inference philosophy is that such conclusions can be made from experiments within the given data base. With much less details about the model we can draw subsets of our data or random samples from the given sample and study our estimate. This means that theoretical investigations for a class of probability measures over the entire sample space are replaced by data base variations. This can only be made efficiently on a computer (unless the data base is very small) and since the experiments will be repeated many times these methods become computer intensive.

Notice that by both classical and new philosophy the end result has to be based entirely on the given data base. This is the only information available for our statistical analysis besides some general information about the situation.

Let us indicate how this works, starting with the classical philosophy. In simple situations of the kind treated in first courses in mathematical statistics one can use analytical methods and tables to construct confidence intervals for parameters. Typical cases are intervals for μ and σ based on samples from the normal distribution.

If $x_1, \ldots, x_n$ is our observed sample of independent variables from the $N(\mu, \sigma)$-distribution, and $\bar{x}, s$ are the estimated mean and standard deviation, then normal distribution theory says that $(\bar{x} - \mu)/(s/\sqrt{n})$ is $t(n-1)$-distributed. This refers to the abstract probability description of all possible samples $x_1, \ldots, x_n$ coming from the normal distribution. (A short exact notation for this is $\frac{\overline{X} - \mu}{s(\mathbf{X})/\sqrt{n}}$, but sometimes the true meaning of this notation has a tendency to pass by unnoticed.) For μ we find the confidence interval $\mu = \bar{x} \pm a\, s/\sqrt{n}$, where a is taken from the tabled $t(n-1)$-distribution so that $P(-a < \frac{\bar{x} - \mu}{s/\sqrt{n}} < a) = c$ is the required confidence level. Similarly $(n-1)s^2/\sigma^2$ has the $\chi^2(n-1)$-distribution and we can find values a and b such that $P(a < \frac{(n-1)s^2}{\sigma^2} < b) = c$, and solve a confidence interval for σ.

Thanks to the central limit theorem and the usually fast convergence of the distributions of averages and other estimates to the normal distribution, or distributions derived from the normal, we can make approximate confidence intervals for the parameters of many other distributions as well. In such cases the classical philosophy is very successful and is often asymptotically optimal.

There is, however, a wide class of problems where the asymptotic distribution is troublesome or where the data are insufficient for useful asymptotic approximations. In a medical application we may for example get many data from each individual. These data may have a complicated dependency. Such data from different individuals can be regarded as independent blocks of observations. An overall measure computed from all the data can be difficult to analyse theoretically, but we can easily vary our data base and notice how the measure varies. If this is made at random, varying which individuals (blocks) we use for the estimate according to rules given later, we get a bootstrap analysis. The bootstrap technique is a method of estimating uncertainty in many such difficult cases. The technique is primarily intended for parameter uncertainty and we describe it in Chapters 5 and 6.

The uncertainty of forecasts can, like parameter uncertainty, be judged by analytical methods in simple standard situations.

If $x_1, \ldots, x_n$ is a sample and we want to predict x_{n+1} by $\hat{x} = \bar{x} = \frac{1}{n}(x_1 + \ldots + x_n)$ then for the corresponding stochastic variables we have $E[X_{n+1} - \hat{X}] = 0$ and $E[(X_{n+1} - \hat{X})^2] = \sigma^2(1 + \frac{1}{n})$. This follows from the i.i.d. (independent identically distributed) assumption usually made for samples, and the assumed existence of a variance. Assuming the data are $N(\mu, \sigma)$, we also know that $X_{n+1} - \hat{X}$ is $N(0, \sigma\sqrt{1 + \frac{1}{n}})$ and that $(X_{n+1} - \hat{X})/s\sqrt{1 + \frac{1}{n}}$ is $t(n-1)$-distributed. Based on the t-distribution we can make precise statements about x_{n+1}, and for example construct a predictive distribution for the corresponding stochastic variable from the information in the sample. However, in more complicated cases one will again come short here. Suppose we can only construct a prediction formula but are unable to make a valid model for the sample. By imitating the prediction situation on the data we already have we can instead assess the prediction uncertainty by systematic cross validation or forward validation. This is described in Chapters 3 and 4.

2.1 Sources of error

It is of course necessary that all the relevant sources of uncertainty are in action in the validation or the bootstrap analysis. Some of them are obvious but others can be more hidden.

We will classify uncertainty and error in four different categories. The first three are

(i)   uncertainty due to the limited set of observations;

(ii)   systematic error due to model defects;

(iii)   further prediction error due to the new residual.

These three sources of error, in particular (i) and (iii), are the ones that are usually considered. Since it may be difficult to formulate a good model on the basis of a priori knowledge alone, one often attempts to counteract systematic model defects, point (ii), by trying many different models on the data. After this experimentation, one of the models is selected according to a criterion which can be based on estimated prediction errors, or the likelihood achieved at the parameter fitting or the like. In such situations we have a further source of uncertainty which can be named

(iv)   model selection uncertainty.

Sometimes, when many persons are involved in data collection and analysis, important decisions about variable selection and models can be based on various preliminary analyses. This is typically done in such a way that no traces of this work can be seen in the data which are taken for final statistical analysis. In such cases one has spoiled the chance to evaluate the effects of the model selection within these data. A typical result is then that uncertainties are underestimated.

2.2 Model selection uncertainty

We will now discuss more carefully why the model selection uncertainty usually will be too complex to be handled analytically. This will serve as one important example of phenomena which require computer intensive statistical techniques.

Consider an experiment leading to a vector of observations and let a sample space Ω represent all possible outcomes of the experiment, possibly extended to allow some extra variables. Suppose we study an estimate $\hat{\theta}$ of a parameter θ or a forecast $\hat{Y}$ of a variable Y. An analytical estimate of uncertainty is usually based on an expected value like $E[(\hat{\theta}-\theta)^2]$, $E[(\hat{Y}-Y)^2]$ or on some probability like $P(|\hat{\theta}-\theta| > c)$.

The expectation is computed as an integral, or sum, over all possible outcomes of the experiment (i.e. over Ω) and with one of the probability measures, for example the one related to the observed value θ_0 of the estimated parameter $\hat{\theta}$. This computation reflects the ideas that the same estimator $\hat{\theta}$ will always be used, and that the probability measure used

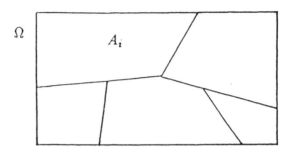

Figure 2.1 *Partitioning of the sample space according to selected models.*

will typically give results close enough to those of the true measure. (The probability $P(|\hat{\theta} - \theta| > c)$ checks $\hat{\theta}$ for every outcome and is basically of the same nature as the expected value.)

Let now $M_1, M_2, \ldots, M_k$ be k parametric models, and let the parameter (vector) θ_m of model M_m belong to a parameter set Ψ_m of possible parameters for this model. When the observations are used to select between the k models, this means that we introduce a *partitioning* of the sample space Ω into disjoint subsets

$$A_i = \{\text{model } M_i \text{ is selected}\}. \qquad (2.1)$$

This partitioning is schematically illustrated in Figure 2.1. and it will introduce two fundamental difficulties. First of all, estimates will be evaluated by different probability measures depending on which set A_i of the sample space our outcome happens to belong to. Secondly, we will in many cases estimate quite different parameters for different A_i since the models can have quite different parameter sets. Analytical methods intended for a given kind of estimate and a fixed model can therefore be entirely misleading. The effect can be seen already in situations with two models but is often more emphasized with many models.

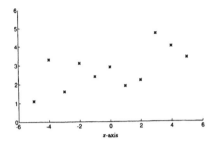

Figure 2.2 *Eleven observations with a possible trend.*

Example 2.1   Selection of models with or without regression.

Eleven independent observations $y_1, \ldots, y_{11}$ are taken. The values are drawn in Figure 2.2 and their positions on the x-axis follow the order in which they were taken. We need an estimate of the expected value for the last observation. Suppose that background information says that there may possibly be a trend in the values, but probably there is none. In order to allow for such a possibility we define $x_i = i - 6$ and choose between the models

$$M_1: \qquad Y_i = \beta_0 + \varepsilon_i,$$

$$M_2: \qquad Y_i = \beta_0 + \beta_1 x_i + \varepsilon_i, \quad i = 1, \ldots, 11;$$

where in both cases ε_i are independent $N(0, \sigma)$.

Our parameter of interest is $\theta = E[Y_{11}]$. This is $\theta = \beta_0$ in model M_1 and $\theta = \beta_0 + 5\beta_1$ in model M_2. A possible procedure in this situation is to estimate β_1 and test if the parameter is significantly different from zero. Set $H_0 : \beta_1 = 0$ and $H_1 : \beta_1 \neq 0$ and suppose we decide to select model M_1 unless H_0 is rejected at the significance level $\alpha = 0.10$. Since $\bar{x} = 0$, simple linear regression gives (for the stochastic version of the estimate) that $\hat{\beta}_1 = \sum x_i Y_i / \sum x_i^2$ is $N(0, \sigma / \sqrt{\sum x_i^2})$ under H_0. The appropriate test rejects H_0 if $|\hat{\beta}_1| / (s_2 / \sqrt{\sum x_i^2}) > 1.833$, where

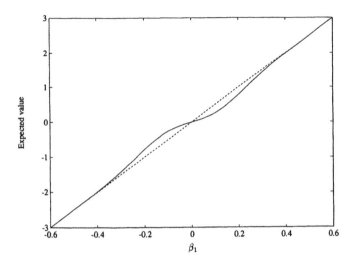

Figure 2.3 *$E[\hat{\theta}_3]$ and true θ as functions of β_1, $\sigma = 1$.*

$s_2 = \sqrt{\sum_{i=1}^{11}(y_i - \bar{y} - \hat{\beta}_1 x_i)^2/9}$ is the M_2-estimate of σ, and 1.833 is the tabulated value in the $t(9)$-distribution.

Let us compare the classical analysis and the result of this two-models procedure. Analysis of one model at a time gives

in M_1 $\hat{\theta}_1 = \bar{y}$

$E[\hat{\theta}_1] = \beta_0$

$\mathrm{Var}(\hat{\theta}_1) = \sigma^2/n = \sigma^2/11;$

in M_2 $\hat{\theta}_2 = \bar{y} + 5\hat{\beta}_1$

$E[\hat{\theta}_2] = \beta_0 + 5\beta_1$

$\mathrm{Var}(\hat{\theta}_2) = \sigma^2/n + 25\sigma^2/\sum x_i^2 = 35\sigma^2/110.$

We have used $\sum x_i^2 = 110$ and $n = 11$.

In each model, the estimate is unbiased and the variance is constant. Model M_2 gives 3.5 times as much variance as model M_1, so it should not be used without reason according to this analysis.

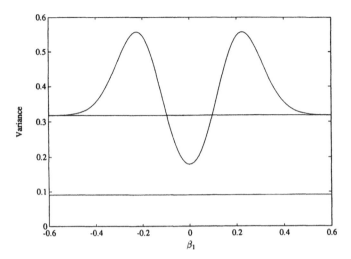

Figure 2.4 *The variance of $\hat{\theta}_3$ compared to variances of $\hat{\theta}_1$ (lower line) and $\hat{\theta}_2$ (upper line) as functions of β_1, $\sigma = 1$.*

The partitioning of the sample space, due to the model selection procedure, gives another estimator defined as

$$\hat{\theta}_3 = \begin{cases} \bar{y} & \text{if } \frac{|\hat{\beta}_1|}{s_2/\sqrt{110}} < 1.833 \\ \bar{y} + 5\hat{\beta}_1 & \text{otherwise.} \end{cases} \tag{2.2}$$

We will give two graphical illustrations of how this estimator works. In Figure 2.3 we can observe that the estimator is no longer unbiased. In Figure 2.4 there is a good swing in the variance when drawn as a function of β_1. Thus, the model selection has changed matters dramatically, and the interesting range of parameter values turns out to be exactly the range where both models have a good chance of being selected.

Our next example treats a somewhat similar situation for a time series. The example is arranged to illustrate the model selection effects. Only the two simplest autoregressive models will be considered.

Example 2.2   Time series models

To an observed stationary time series $y_1, \ldots, y_n$ with mean zero, autoregressive models of the first and second order are fitted.

$$M_1 : \quad Y_t = cY_{t-1} + \varepsilon_t \qquad\qquad (2.3a)$$

$$M_2 : \quad Y_t = aY_{t-1} + bY_{t-2} + \varepsilon_t \qquad (2.3b)$$

The deviations ε_t are independent of each other, and each ε_t is also independent of previous values $Y_{t-1}, Y_{t-2}, \ldots$ in the models of the time series. Let $E[\varepsilon_t] = 0$ in both models and $\mathrm{Var}(\varepsilon_t) = \sigma_1^2$ in model M_1 and σ_2^2 in model M_2. Let our models be stationary and define covariances $R_i = E[Y_{t+i}Y_t]$ at the time distance i. By this notation $\mathrm{Var}(Y_t) = R_0$. The corresponding correlations are then $\rho_i = R_i/R_0$. From our data we estimate R_0, R_1, R_2 and ρ_1, ρ_2 by the following model independent formulas

$$\hat{R}_0 = \frac{1}{n} \sum_{t=1}^{n} y_t^2 \; ; \quad \hat{R}_i = \frac{1}{n} \sum_{t=1}^{n-i} y_{t+i} y_t \; ; \quad \hat{\rho}_i = \hat{R}_i / \hat{R}_0.$$

Next we derive estimates of σ_1 and σ_2. If we multiply the model M_1 (2.3a) by Y_t and take expected values of each term and then repeat the same thing with Y_{t-1} as multiplier instead of Y_t we get the equations

$$\begin{cases} R_0 = c\,R_1 + \sigma_1^2 \\ R_1 = c\,R_0 \, . \end{cases}$$

Solving for σ_1^2 we get

$$\sigma_1^2 = R_0(1 - \rho_1^2). \qquad\qquad (2.4)$$

In the same way we can multiply the model M_2 with Y_t, Y_{t-1}, and Y_{t-2} respectively. Taking expected values for each multiplier we get the equations

$$\begin{cases} R_0 = a\ R_1 + b\ R_2 + \sigma_2^2 \\ R_1 = a\ R_0 + b\ R_1 \\ R_2 = a\ R_1 + b\ R_0. \end{cases}$$

After elimination of a and b and simplification we get

$$\sigma_2^2 = R_0(1 - \rho_1^2)\left(1 - \frac{(\rho_2 - \rho_1^2)^2}{(1 - \rho_1^2)^2}\right). \qquad (2.5)$$

The variances σ_1^2 and σ_2^2 can be interpreted as the one step ahead prediction errors, if the models M_1 or M_2 are correct. If M_1 is true, one can show that $\rho_2 = \rho_1^2$. The two expressions are then equal. When $\rho_2 \neq \rho_1^2$ the expression (2.5) is always smaller. In spite of this, it does not follow that M_2 will be better than M_1 when estimation errors are taken into account. On what basis should we select one of them? Let $\hat{\sigma}_p^2$, $p = 1, 2$, be the estimated variances when $\hat{R}_0$, $\hat{\rho}_1$, $\hat{\rho}_2$ is put into (2.4), respectively (2.5). According to a simple model selection criterion (forward prediction error, FPE) suggested by Akaike (1969), $\hat{\sigma}_p^2$ is a potential underestimation due to the overfitting of p parameters to the data. This overfit is compensated for by multiplying the estimated prediction error variance by a factor $(1 + p/n)/(1 - p/n)$. (The factor is derived by an asymptotic argument.) According to Akaike's suggestion we will therefore compare $V_1 = \hat{\sigma}_1^2(1 + 1/n)/(1 - 1/n)$ with $V_2 = \hat{\sigma}_2^2(1 + 2/n)/(1 - 2/n)$ and select the model with the smallest estimated prediction error after this correction. Let us see what this procedure means.

Divide V_1 by V_2 using (2.4) and (2.5). We will then find that by this criterion M_2 should be used if

$$1 - \frac{(\hat{\rho}_2 - \hat{\rho}_1^2)^2}{(1 - \rho_1^2)^2} < \frac{(1 - \frac{2}{n})(1 + \frac{1}{n})}{(1 + \frac{2}{n})(1 - \frac{1}{n})} = 1 - \frac{\frac{2}{n}}{(1 + \frac{2}{n})(1 - \frac{1}{n})}$$

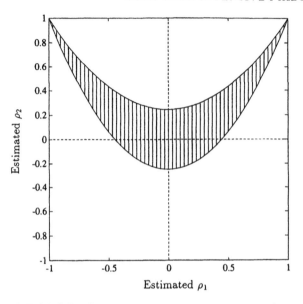

Figure 2.5 *Model selection areas for time series with 32 observations and the FPE criterion.*

that is if

$$\left| \frac{\hat{\rho}_2 - \hat{\rho}_1^2}{1 - \hat{\rho}_1^2} \right| > \sqrt{\frac{2/n}{(1 + \frac{2}{n})(1 - \frac{1}{n})}}.$$

The special case $n = 32$ is illustrated by a $(\hat{\rho}_1, \hat{\rho}_2)$-diagram in Figure 2.5. In the shadowed area of the figure we select model M_1, outside this area we take model M_2.

A typical sample space for this problem will be of high dimension and have at least 32 dimensions. It is therefore not useful for illustrations. The partitioning in Figure 2.5 corresponds, however, to a similar partitioning of the sample space.

This partitioning has some unpleasant consequences. An autoregressive model of order two with its true (ρ_1, ρ_2) inside the shaded area will for example seldom be correctly selected and if selected, the estimate $(\hat{\rho}_1, \hat{\rho}_2)$ can never take values in a neighbourhood of the true parameter point. Just like in Example 2.1, expected values and variances etc. computed without considering model changes can be quite meaning-less or at best approximations. Also the analysis behind the model selection criterion and the factors $(1 + p/n)/(1 - p/n)$ can be questioned.

By computer intensive techniques, model selection effects can be considered. The trick is to let the model selection be part of the inference procedure and therefore variable when we vary our data base.

Concluding remarks

The classical philosophy, based on sample space models, and the computer intensive philosophy, based on data base varia-tions, are complementary and not in conflict with each other. It is natural to handle classical well structured problems by the traditional methods, and many complex and less struc-tured problems by the computer intensive methods. Model selection is just one example, but an important one, of prob-lems which can be approached by the new methods. Virtu-ally anything calculated from uncertain data can be of in-terest. Classification, image analysis, interpretation of satel-lite data, time series problems, factor analysis, etc. are all within the range of interest for the new methods. But the computer intensive philosophy stands on the shoulders of the traditional philosophy. We learn its properties on classical examples and with classical evaluation methods before we extrapolate into more difficult and unknown situations.

CHAPTER 3

Cross validation

3.1 Introduction

Cross validation is a method of evaluating given models by means of their forecasts and to choose a model with proper complexity. In a series of examples we will first show how this works in a very elementary situation, which is later developed into a more complex analysis. To simplify the presentation our first examples will treat rather small data sets and model sets. This should not be misunderstood. Cross validation is perhaps most valuable when the data are highly multivariate, or complex in some other way, and many models are conceivable. However, larger analyses are best handled directly on a computer. We will demonstrate that cross validation can be performed in different ways, and the main message of this chapter is the cross model validation method we suggest as an instrument in model selection problems.

Standing notation in this chapter is:

DS = data set = all data which can be analysed.
ES = estimation set = subset of DS used for estimation.
TS = test set = subset of DS used for testing.

If $\hat{y}$ is a prediction of y, let $L(y, \hat{y})$ denote the loss function. When y and $\hat{y}$ are scalar, we are usually interested in the squared error

$$L(y, \hat{y}) = (y - \hat{y})^2, \tag{3.1}$$

but also $|y - \hat{y}|$ and more special measures (compare (4.11)–(4.14)) can be of interest.

As an average of the loss function (over predictions of n variables $y_1, \ldots, y_n$) we will often use the symbol

$$Q = \frac{1}{n} \sum_{i=1}^{n} L(y_i, \hat{y}_i). \qquad (3.2)$$

When necessary we use indices on these symbols to distinguish different versions. In theoretical studies we can take expected values in a model and for example introduce $L(Y, \hat{Y}) = E[(Y - \hat{Y})^2]$. The measures we compute on the data can then be regarded as estimates of such expected values. When we want to distinguish between these measures, we will write L and Q for the theoretical measures and $\hat{L}$ and $\hat{Q}$ for the computed estimates.

Example 3.1
A data set DS is divided into two parts, the estimation set ES and the test set TS.

Estimation set:

x	2	5	12	4	6	7	8	10	12
y	22	5	7	-6	46	2	13	25	22

Test set:

x	5	8	10	6	6	7	13	15	7	7
y	10	8	25	-14	-29	37	28	0	-13	27

Each item of data has two values (x, y) and we want to predict new y-values after the corresponding x-values are known. A prediction model $\hat{y} = ax$ is studied using the square loss function (3.1). This is the simplest possible prediction model for the situation, but it has the basic properties of a more general regression model. As an overall measure we use the mean square error $Q = \frac{1}{n}\Sigma(y_i - \hat{y}_i)^2$. The parameter a is estimated by regression on ES. This gives

$$\hat{a}_{ES} = \frac{\Sigma x_i y_i}{\Sigma x_i^2} = 1.78. \qquad (3.3)$$

The corresponding minimal Q-value on ES becomes

$$Q_{ES} = 238.$$

The formula $\hat{y} = \hat{a}_{ES}x = 1.78x$ can now be tried on TS. This gives the much higher mean square error

$$Q_{TS} = 448.$$

We may perhaps also compute $\hat{a}_{TS} = 1.03$ and compare with $\hat{a}_{ES}$.

Discussion

The example shows a very simple version of model testing, which has often been used but mostly in more complex situations with much more extensive models. One can study this testing method critically and ask whether any knowledge was achieved from TS which could not have been achieved directly from ES or from a direct analysis of the entire DS.

Within the conventional theory of regression analysis, assuming normal distribution,

$$Y = ax + \varepsilon, \quad \varepsilon \text{ independent } N(0, \sigma) \qquad (3.4)$$

all the *statistical* properties of $\hat{a}$ and of the forecasts can be derived without TS. We know for example that $\hat{a}$ is $N(a, \sigma/\sqrt{\Sigma x_i^2})$; $E\hat{Q}_{ES} = \sigma^2(1 - 1/n)$. (Summations and n are for the data used in the analysis, e.g. ES. The notation $\hat{Q}$ is used here since we are in a model situation.)

The test on the new data, TS, can give some information about the estimation error $\hat{a} - a$ and about σ^2. Theoretical arguments show, however, that this information can never exceed the information from a direct analysis of the entire normally distributed sample DS. It can be shown that $\hat{a}_{DS}$ and $s_{DS}^2 = n\hat{Q}_{DS}/(n-1)$ are sufficient for the model (3.4). This means that they contain all the available information about the parameters a, σ^2. The two-step procedure of Example 3.1 can therefore only make sense if the normal model is wrong or the test set TS for some reason is not available in the estimation.

Figure 3.1 *Take one out*

It is left to the reader to consider what conclusions we may draw from the relatively large differences in $\hat{a}$ and Q between ES and TS. Are further analyses needed, and if so, are we better off afterwards.

Example 3.2
Let DS $= \{(x_i, y_i),\ i = 1, 19\}$ consist of all data of the preceeding example. A more efficient way to evaluate a prediction formula was suggested by Allen (1971) and given the name PRESS = prediction sum of squares. The idea is illustrated in Figure 3.1. Partition DS in ES_j and TS_j, where ES_j = all data except the jth and $TS_j = (x_j, y_j)$. In our data the set ES_j contains 18 observations and TS_j the nineteenth.

We will again use the prediction formula $\hat{y} = ax$ and the loss function $L(y_j, \hat{y}_j) = (y_j - \hat{y}_j)^2$ as in Example 3.1. Let $\hat{a}_j$ be estimated on ES_j by the formula

$$\hat{a}_j = \frac{\sum_{i \neq j} x_i y_i}{\sum_{i \neq j} x_i^2} \qquad (3.5)$$

in analogy with (3.3). The values for $j = 1, \ldots, 19$ are computed as

    1.32 1.35 1.44 1.38 1.18 1.39 1.33 1.26 1.29 1.33
    1.36 1.26 1.45 1.51 1.20 1.23 1.61 1.46 1.25

The sum over j of the corresponding losses is called prediction sum of squares and denoted PRESS, and for the average we define the symbol

$$Q_{\mathrm{CV}} = \frac{\mathrm{PRESS}}{n} = \frac{1}{n} \sum_{j=1}^{n} L(y_j, \hat{y}_j) \qquad (3.6)$$

or more explicitly $\frac{1}{n} \sum (y_j - \hat{a}_j x_j)^2$. The index CV is short for cross validation. We get the result $Q_{\mathrm{CV}} = 365$. As a final prediction formula we compute on the entire DS

$$\hat{a} = \frac{\sum x_i y_i}{\sum x_i^2} = 1.35$$

and by this method $\hat{y} = 1.35x$ will be used on future observations.

Discussion

The new element compared to Example 3.1 is that all the data are used for the evaluation of the predictability. We have also departed from the idea that a unique prediction formula should be validated since $\hat{a}_j$ varies slightly with j. Instead we evaluate the *method* to compute a prediction formula (the final one). The Q_{CV}-measure gives an estimate of the average loss, but can in some cases only be considered relevant for an average over the x-values in the data base DS.

The computer intensive element in the example is that the regression coefficient $\hat{a}$ is computed $n + 1$ times instead of just once. Further comments on the use of this method in model selection will be given in Section 3.4, see also Section 3.8.

3.2   Cross validation as an estimation method

A new element in cross validation was introduced by Stone (1974). He suggested that some parameters should be optimized with regard to the cross validation measures. Applying this idea to the same data DS as in the examples above, we can use the approach by Stone and for example introduce a parameter α so that

$$\hat{y}(\alpha) = \alpha \hat{a} x. \qquad (3.7)$$

Here $\hat{a}$ is computed as in Examples 3.1 and 3.2, and α is taken as a shrinking parameter. It may seem overparametrized to introduce a product $\alpha \hat{a}$, but in the cross validation we will keep α fixed and let $\hat{a}$ vary according to which data ES we use for the estimation.

Example 3.3
Let $\hat{a}_j$ be computed as in (3.5) for each j. Put

$$Q_{\mathrm{CV}}(\alpha) = \frac{1}{n} \sum_{j=1}^{n} (y_j - \alpha \hat{a}_j x_j)^2$$

and minimize this with respect to α. Differentiate this to get the equation

$$Q'_{\mathrm{CV}}(\alpha) = \frac{1}{n} \Sigma 2(y_j - \alpha \hat{a}_j x_j)(-\hat{a}_j x_j) = 0.$$

From this we solve

$$\hat{\alpha} = \frac{\Sigma y_j (\hat{a}_j x_j)}{\Sigma (\hat{a}_j x_j)^2} = 0.87 \qquad (3.8)$$

which gives a minimum $Q_{\mathrm{CV}}(0.87) = 363$.
    The final prediction formula $\hat{y} = 1.35x$ will now be replaced by $\hat{y} = 0.87 \times 1.35x = 1.17x$.

Discussion of Example 3.3
We find that an estimated value $\hat{\alpha} < 1$ becomes optimal. This means that the regression coefficient will be adjusted down towards zero to give the best forecasts according to the measure Q_{CV}. This is a known property, which can also be achieved by more analytical methods. It is associated with the famous results by Stein (1956) and James and Stein (1961). We give up the unbiasedness of the estimated regression coefficient in order to get smaller prediction errors. However, in situations with few parameters the effect

should not be exaggerated. We can compare the minimum $Q_{CV}(0.87) = 363$ with $Q_{CV}(1) = 365$ computed by (3.6).

Notice how easily the parameter α is estimated by cross validation. An analytical determination would require both extensive definitions and tricky computations.

In the three examples 3.1–3.3 we have worked with the simplest possible regression equation in order to demonstrate some ideas with a minimum of computations. Of course the same ideas can be applied to larger models. Often the simplest useful model will include a constant term such as in the model $Y = \beta_0 + \beta_1 x + \varepsilon$. We may then work with a shrinking factor $\alpha\hat{\beta}_1$ just as in Example 3.3, but will typically not shrink the constant β_0. We can go on to multiple regression and shrink the entire equation uniformly (except for the constant) as in Copas (1983) or shrink in some more sophisticated manner. A useful application of shrinking to traffic accident data was made by Junghard (1990). This is described further in Chapter 7. Let us here only mention that in this particular application the shrinking turned out to be necessary for correct estimates, when crosses with many accidents were selected for modification.

The shrinking parameter α in Example 3.3 can be replaced by much more general parameters. We can for example let α vary over a set which contains several different prediction models and choose the one which minimizes the cross validation measure. However, in order to discuss this we need a more extensive set of data.

3.3 Selection of variables in multiple regression

In the statistical literature several different procedures have been suggested for variable selection in regression. The most well known are based on best subset regression, forward selection, and backward elimination, together with some decision rule for the model size. See e.g. Hocking (1976) for a review of this tradition. The problems considered in the later sections of this chapter are not solved by these older approaches. Another suggested approach is to not select,

Table 3.1 *Cement hardening data, the 'Hald data'; amounts x_i in percentages of weight, heat in calories per gram.*

x_1 3CaO Al$_2$O$_3$	x_2 3CaO SiO$_2$	x_3 4CaO Al$_2$O$_3$ Fe$_2$O$_3$	x_4 2CaO SiO$_2$	y heat
7	26	6	60	78.5
1	29	15	52	74.3
11	56	8	20	104.3
11	31	8	47	87.6
7	52	6	33	95.9
11	55	9	22	109.2
3	71	17	6	102.7
1	31	22	44	72.5
2	54	18	22	93.1
21	47	4	26	115.9
1	40	23	34	83.8
11	66	9	12	113.3
10	68	8	12	109.4

but let all variables enter into the model and then shrink the equation with a proper shrinking factor α as in Example 3.3. See Copas (1983). We will not treat such shrinking factors here.

If we only have a few predictors, which are not too close to linear dependency, we may in many cases be best off by bringing them all into the model. When the number of predictors increases, we may gain a lot by selecting smaller models than the maximal one, and sometimes it is quite necessary to exclude predictors (linearly dependent predictors, more predictors than observations, or practical arguments for small models). As an example we will consider the data in Table 3.1, which have been analysed by several authors before: Hald (1952), Draper and Smith (1981) and Stone and Brooks (1990) to mention just a few. The figures are about the hardening of cement and the heat evolved during the first 180 days after addition of water.

In total we may fit $2^4 - 1 = 15$ different models

Table 3.2 *Measures Q_{CV} and s^2 for the 15 possible regression models.*

Variables in the model				Q_{CV}	s^2
x_1	x_2	x_3	x_4		
1	2	3	4	8.49	5.99
1	2	3		6.92	5.35
1	2		4	6.57	5.33
1		3	4	7.27	5.65
1	2			7.22	5.79
1		3		170.62	122.71
1			4	9.33	7.48
1				130.74	115.06
	2	3	4	11.30	8.20
	2	3		53.48	41.54
	2		4	112.45	86.89
	2			92.47	82.39
		3	4	22.62	17.57
		3		201.26	176.31
			4	91.86	80.35

$$y = \beta_0 + \sum_{i=1}^{p-1} \beta_{j_i} x_{j_i} + \varepsilon$$

containing at least one of the x-variables. For each such model the measure Q_{CV} in Example 3.2 can be computed. See formula (3.6). As comparison also the measure s^2 can be computed, where $s^2 = \text{RSS}/(n-p)$ and RSS is the residual sum of squares in ordinary regression, $n = 13$ is the number of data items, and p is the number of β-parameters including the constant β_0.

It is practical, especially with large sets of models, to organize the results according to the model sizes and only keep the best model of each size, according to the measure. Doing so, we will also prepare our notations for a more theoretical

Table 3.3 *The best model of each model size.*

Model size	Best model for $\hat{y}$	$\hat{Q}_{\mathrm{CV}}(p)$
$p=2$	$117.6 - 0.74x_4$	91.86
$p=3$	$52.6 + 1.47x_1 + 0.66x_2$	7.22
$p=4$	$71.6 + 1.45x_1 + 0.42x_2 - 0.24x_4$	6.57
$p=5$	$62.4 + 1.55x_1 + 0.51x_2 + 0.10x_3 - 0.144x_4$	8.49

discussion later. We will therefore start talking about true and estimated measures.

Let $\hat{Q}_{\mathrm{CV}}(p)$ be the estimated measure Q_{CV} for the best model of size p, let $\hat{p}$ be the best model size, and $\hat{Q} = \hat{Q}_{\mathrm{CV}}(\hat{p})$ the best estimated measure (over all the models).

$\hat{Q}_{\mathrm{CV}}(p) = \min \{Q_{\mathrm{CV}}, \text{models with } p \text{ parameters}\},$

$\hat{p} = \{p \text{ such that } \hat{Q}_{\mathrm{CV}}(p) < Q_{\mathrm{CV}}(p') \text{ all } p' \neq p\},$

$\hat{Q} = \hat{Q}_{\mathrm{CV}}(\hat{p}) = \min \{\hat{Q}_{\mathrm{CV}}(p)\}.$

This formulation supposes a unique best size $\hat{p}$, otherwise let $\hat{p} = \min \{p \text{ such that } \hat{Q}_{\mathrm{CV}}(p) \leq \hat{Q}_{\mathrm{CV}}(p') \text{ all } p'\}$ be the smallest of the best model sizes.

The final selection is between these models and provided no external arguments, such as costs, interact in the decision, we will by this method select

$$\hat{p} = 4, \quad \hat{Q} = 6.57, \quad \text{the model above.}$$

When the number of models becomes very large, we cannot for practical reasons study each model any longer. Instead we usually prefer some standard procedure for stepwise regression, which automatically picks one model for each model size p based on the comparison of a more limited number of models. We will comment more on this later.

3.4 A theoretical difficulty and a deeper look at the model selection measure

Are the measures $\hat{Q}_{\text{CV}}(p)$ and $\hat{Q}$ in Section 3.3 good estimates of the predictability for the corresponding models?
These measures do have a good intuitive basis since we have
taken the averages of all prediction losses (3.6) for the selected models. We will here demonstrate that, nevertheless,
they are biased and too low *due to the model selection itself.*
In order to show this we first need a simple result.

Proposition   If $X \geq 0$ and $P(X = 0) < 1$ then $EX > 0$.

Proof   Although 'obvious', it can be difficult to write down
a correct proof. Start with

$$EX = \int_0^\infty x\, dF(x) \geq \int_a^\infty x\, dF(x) \geq aP(X \geq a) \qquad (3.9)$$

which is valid for any $a \geq 0$.
   Let a_i be a positive sequence of real numbers decreasing
towards 0, ($a_i = 1/i$ for example). Then we have

$$P(X > 0) = P(\cup_{i=1}^\infty \{X \geq a_i\}) \leq \Sigma P(X \geq a_i),$$

since $P(\cup A_i) \leq \Sigma P(A_i)$ is always valid. If $P(X \geq a_i) = 0$
for all i then $P(X > 0) = 0$ and $P(X = 0) = 1$ contradicting
the proposition's condition. Thus $P(X \geq a_i) > 0$ must be
true for some i, and (3.9) gives the result.

   With this proposition we will now show that the model selection causes underestimation of the average prediction loss.
The following situation is considered: an observation vector
$Y = (y_1, \ldots, y_n)'$ is regarded as an observed random vector
from a probability space with a true probability measure P.
Let E denote expected value computed with the true measure P. Every such expected value is of course computed as
an integral over the entire sample space.

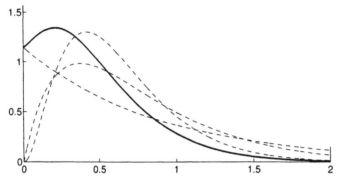

Figure 3.2 *Densities of unbiased preformance estimates for three regression models, dashed, and for the selected minimum measure which is biased, solid.*

To the data we fit different models M_i, and $\hat{Q}_i$ is an estimated measure for the model M_i. The model M_j which minimizes the estimated loss will be selected.

Theorem 3.1   Selection bias
Let $\mu_i = E\hat{Q}_i$ and $\hat{Q} = \min \hat{Q}_i$. Then

$$E\hat{Q} \leq \min \mu_i \leq \mu_j$$

for every j including the j corresponding to the selected model. The inequalities are strict unless $\hat{Q} = \hat{Q}_j$ with probability 1 for some j.

Proof Define a variable $X = \hat{Q}_i - \hat{Q} \geq 0$ for every outcome. The proposition gives $EX = E\hat{Q}_i - E\hat{Q} > 0$, unless $\hat{Q}_i = \hat{Q}$ with probability 1 when there is equality. The first inequality is then valid for every i including the one corresponding to min μ_i. The second inequality is trivially true.

It is natural to look for an optimal model minimizing some loss function (or maximizing some utility function). The main point of the theorem above is that the very selection of such a model introduces bias error in the measure. As stated before the model choice introduces a partitioning of the sample space Ω into events $A_j = \{\hat{Q} = \hat{Q}_j = \min \hat{Q}_i\}$ defined as the set of samples where the model M_j is selected. It is this partitioning which is critical for the statistical properties when we want to study many different models and choose one of them on the basis of observed data. One can study other effects of the partitioning and find that not only does $\hat{Q}$ underestimate the expected value for each model, but there may also be an opposite effect for the true measure of the selected model. If a true model competes all the time with other models and we increase the number of competing models, the measure $\hat{Q}$ can only decrease. However, the ability to predict will usually become worse due to overfitting of the data and since the true model is selected by a smaller and smaller probability. One has to be prepared to pay something for the search among many suggested models, but this cost is impossible to estimate by conventional statistical methods.

On the other hand, if the best model is not there to begin with, the (expected) true predictability of the selected model may improve when the number of models increases. It is therefore important to have good estimates of the true predictability. An example of model selection effects in regression analysis is shown in Figure 3.3. The corresponding illustration with 20 predictors is given in Hjorth (1989).

Notice the very strong effects of the model selection in Figure 3.3 on the ratio between the mean estimated criterion and mean value of the true criterion. The case with ten potential predictors and 40 data points is not at all extreme. However, the situation $\beta = 0$ is seldom realistic. What happens when $\beta \neq 0$? It is illustrated in Hjorth (1989) that for β slightly different from zero the bias of the estimated criterion can become even worse. However, if a component β_k is very large, compared to the standard deviation of its estimate, it will almost always be selected before the small components. This large parameter will therefore not contribute much to the model selection effects. The bias effects illustrated in Figure 3.3 will then refer only to models which contain this obviously necessary parameter and will therefore start at model size 2. In the same way more than one large component can

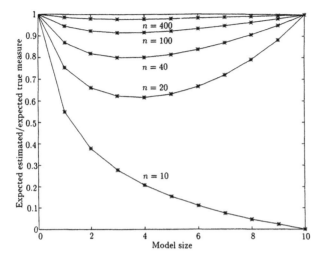

Figure 3.3 *Quotient, due to model selection, between the expected value of the estimated measure and the expected value of the true measure. Each estimate is unbiased if no model selection is involved. Regression is with n observations and 10 orthogonal potential predictors. The best fitting model of each size p is selected. True $\beta = 0$.*

delay the bias of estimated measures to start at larger model sizes.

Model selection bias in regression is not a new or isolated discovery. It has also been observed for example by Rencher and Pun (1980) and by Miller (1990). Recently, a good illustration of model selection effects was given by Breiman and Spector (1992) and Breiman (1992), based on an extensive set of simulations. This knowledge has however, not affected the computer packages yet as far as we know, nor has it been been noticed by the majority of statisticians using stepwise regression and similar tools.

Our results in this section motivate the following question. Is it possible to arrange cross validation so that the model selection effects are measured and $\hat{Q}$ will have a correct expected value? The answer is an approximate yes. A necessary condition for doing this is, however, that one abandons the measure PRESS and related measures where one model

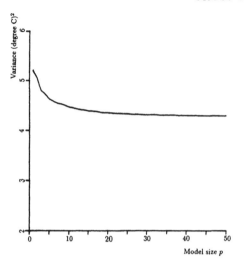

Figure 3.4 $s^2 = \text{RSS}/(n - p)$ *for different model sizes p. Forward selection among predictors. Meteorological data.*

at a time is studied. In order to measure model selection effects by validation, model selection errors must be in action during the analysis. This means that we have to let different models be selected, depending on which data set ES_j we are analysing for the moment. In the next section we give some examples.

3.5 Validation of model size

The problem of selecting a model size is illustrated by two figures. In Figure 3.4 we show the sequence of s^2-estimates, $s^2 = \text{RSS}/(n-p)$, from stepwise regression (forward selection without F-tests) on a set of meteorological data. In this example we have $m = 50$ potential predictors and $n = 1392$ observations. Figure 3.4 gives no good indication of where the analysis should be interrupted.

In Figure 3.5 we show instead the result of a cross validation with built in estimation of the model selection effect (as described below). Here a rather distinct minimum is found at the model size $p_0 = 3$. Since the cross validation is built on copies of typical prediction situations, one can from the construction directly state that Figure 3.5 gives an unbiased

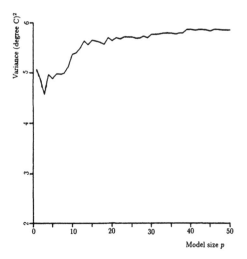

Figure 3.5 *Cross validation measure* CMV(p) *for different model sizes* p. *The same data and forward selection routine as in Figure 3.4.*

picture of the predictability for different p-values (in average over the predictor combination in the data and with $n - 1$ observations in the estimation sets instead of n).

Due to the time dependency of weather, the data will correspond to much fewer independent observations. The cross validation of Figure 3.5 was in this case performed on eight independent data sets taken in different years (Hjorth and Holmqvist, 1981).

We may here, like in Table 3.3, regard the model selection as a two-step procedure. In step one a model is selected for each model size p, and in step two a proper model size p_0 is chosen. In order to compare different models with the same number p of parameters, we may use a proper measure. In regression we can for example take one of (a) to (e) below.

(a)   RSS, the residual sum of squares
(b) $s^2 = \text{RSS}/(n - p)$
(c) $s^2(1 + p/n)$ 'the forward prediction error'
(d) R or R^2, the multiple correlation
(e)   PRESS or Q_{CV}.

The measures (a), (b), (c) and (d) are equivalent if p is constant, but behave differently when different p-values are compared. The measure (d) should be maximized, the others minimized. The measure (c) can for *one* true model be interpreted as an estimate of the mean square error of forecasts of new measurements with exactly the same predictors as in the data set. This statement is based on the following result.

Let $\mathbf{Y} = X\beta + \varepsilon$ and $\mathbf{Z} = X\beta + \nu$ where ε and ν are independent $N(\mathbf{0}, \sigma^2 I)$. The parameter vector β is estimated from $\mathbf{Y}$ as $\hat{\beta} = (X'X)^{-1}X'\mathbf{Y}$ and is used for the prediction of $\mathbf{Z}$. Then

$$\frac{1}{n}E[(\mathbf{Z} - X\hat{\beta})'(\mathbf{Z} - X\hat{\beta})] = \sigma^2(1 + p/n).$$

See Miller (1990, p. 182) or derive the result from the independence of $\mathbf{Z}$ and $\hat{\beta}$; compare Chapter 4 of Rao (1973). The cross validation measure e is about n times as heavy to compute if no special tricks are used. Its interpretation as a forecasting measure was discussed earlier.

3.6 Cross Model Validation

The model selection measure we choose, for example among (a) to (e) above, will now be used as an element inside a cross validation. We call it the inner criterion in the flow chart below. Let $DS = \{(y_1, \mathbf{x}_1), \ldots, (y_n, \mathbf{x}_n)\}$ be our data set. For each j between 1 and n, DS is divided into $TS_j = (y_j, \mathbf{x}_j)$ and $ES_j = DS - TS_j$. The important step is to let the model selection work anew for each ES_j using this measure.

Suppose we have excluded data j. Based on an analysis of ES_j we will then select one model for each model size p. To this end we compute and compare the model selection measure (the inner criterion) for a set of candidate models of the same size p and pick the best. Let $\widehat{M}(p, ES_j)$ be the selected model.

When the model set is reasonably small, all models of size p can be compared and we get a best subset regression. However, sometimes we have large model sets or prefer to limit

the comparison for other reasons. We may then define the
sets recursivily, so that the selected $\widehat{M}(p, \mathrm{ES}_j)$ determines
the set of compared models for $p + 1$. In forward selection
we will for example compare all models with one predictor
added to those of $\widehat{M}(p, \mathrm{ES}_j)$. (We think of the basic forward selection without testing for inclusion or deletion of
variables.) Backward elimination (Draper and Smith 1981)
is another possible procedure if the full model is estimable
on every ES_j.

In order to emphasize the model variation with j, we will
now define a new concept, Cross Model Validation, for this
procedure where

$$\hat{y}_j(p) = \hat{y}_j(\mathbf{x}_j, \widehat{M}(p, \mathrm{ES}_j), \mathrm{ES}_j) \qquad (3.10)$$

is the forecast produced by the model $\widehat{M}(p, \mathrm{ES}_j)$ estimated
on ES_j and

$$\mathrm{CMV}(p) = \frac{1}{n} \sum_{j=1}^{n} (\hat{y}_j(p) - y_j)^2. \qquad (3.11)$$

The squared error can of course be generalized to any loss
function $L(\hat{y}_j(p), y_j)$.

The model size p_0 minimizing $\mathrm{CMV}(p)$ is estimated as optimal and a new and final selection among models of this
size is performed based on all data in DS. Denote this final
model by $\widehat{M}(p_0, \mathrm{DS})$. The computations are illustrated by a
flow chart on the next page.

3.6.1 Flow chart of Cross Model Validation for independent observations

LOOP OVER ALL DATA
   $j = 1, \ldots, n$
   $\text{TS}_j = (y_j, \mathbf{x}_j)$ define test set
   $\text{ES}_j = \text{DS} - \text{TS}_j$ estimation set
   LOOP OVER MODEL SIZES
      $p = 1, \ldots, p_{\max}$
      LOOP OVER MODELS OF SIZE p
         $r = 1, \ldots, m_p$
         $C(M_r(p), \text{ES}_j)$ inner criterion for mod-
         end of r-loop   el $M_r(p)$, data ES_j
      INTERMEDIATE RESULTS
      $\widehat{M}(p, \text{ES}_j)$ best model of size p
                                                 by inner criterion
      $\hat{y}_j(p) = \hat{y}_j(\mathbf{x}_j, \widehat{M}(p, \text{ES}_j), \text{ES}_j)$ prediction
      $\text{CMV}(p) = \text{CMV}(p) + \frac{1}{n}(\hat{y}_j(p) - y_j)^2$ adding outer criterion
      end of p-loop
   end of loop over data
SELECTION OF MODEL SIZE
$p_0;\ \text{CMV}(p_0) \leq \text{CMV}(p)$, all p

NEW LOOP OVER MODELS OF SIZE p_0
   $r = 1, \ldots, m_{p_0}$
   $C(M_r(p), \text{DS})$ repeat on all data
   end
OUTPUT
$\widehat{M}(p_0, \text{DS})$ selected model
$\hat{y}(\mathbf{x}, \widehat{M}(p_0, \text{DS}), \text{DS})$ prediction formula
$\text{CMV}(p_0)$ mean loss

Table 3.4 *Effect of model variation.*

p	CMV(p)	$\hat{Q}_{CV}(p)$
2	108.90	91.86
3	9.72	7.22
4	6.99	6.57
5	8.49	8.49

Example 3.4   The 'Hald data' again.
The cross model validation, CMV(p), is applied to the data of Table 3.1. As comparison we also give the earlier results, $\hat{Q}_{CV}(p)$, of cross validation without model variation.

The selected model has $p_0 = 4$ and is estimated on DS as $\hat{y} = 71.65 + 1.45x_1 + 0.42x_2 - 0.24x_4$.

Comments on Example 3.4
For $p = 2, 3, 4$ the measure CMV(p) is larger than $\hat{Q}_{CV}(p)$. This depends on the model selection acting on CMV(p), since there is no other difference in the measures. The model selection introduces uncertainty in exactly the same manner as our final selection of a $\hat{y}$-model has uncertainty. For $p = 5$ we have just one model and the two measures become identical. In this small example we make the same model choice from the two validation measures, but the estimated mean square prediction errors become different. In most large examples the decisions will also be different, often considerably different depending on whether the model selection effects are allowed to affect the measures or not.

3.6.2   Rationalizations

When the data sets are very large, with hundreds of observations or more, it can be useful to pick several observations at a time in TS. We may for example let ES have 95% of the

Table 3.5 *The number of models in total and of size*
p for different numbers, k, of potential predictors.

	$k = 5$	$k = 10$	$k = 20$	$k = 30$
Total	31	1 023	1 048 575	1.07×10^9
$p = 1$	5	10	20	30
$p = 2$	10	45	190	435
$p = 3$	10	120	1 140	4 060
$p = 4$	5	210	2 845	27 405
$p = 5$	1	252	15 504	142 506
		$\vdots$	$\vdots$	$\vdots$

data and TS 5%. This makes the heavy outer loop over the
data take only 20 turns instead of n.

If the set of potential predictors becomes large, we get a
large model set. With k predictors we have $2^k - 1$ non-empty
models. Of these $\binom{k}{p} = \frac{k!}{p!(k-p)!}$ are of size p. As demonstrated in Table 3.5 the number of models grows rapidly
with k.

With so many models, an analysis of them all can often
be out of the question. It is then usually replaced by a stepwise regression procedure such as Forward selection (FS). In
the CMV flow chart this means that the two cross validation loops over all model sizes, and over models with the
same size p, are put together into a single one since a FS-
procedure or an alternative stepwise procedure will just provide one suggested model of each size p. A search over a set
of models with the same p is instead performed inside the FS-
procedure. Standard packages for stepwise regression often
offer a stop criterion based on F-tests. However, the theoretical basis of this F-test is not at all valid in the multiple
model situation, which makes this stop criterion quite arbitrary and useless in our application. By defining the F-test
criterion to be zero, the FS-routines will run unaffected.

Simulations and experiments on data sets have demonstrated that it is not at all certain that a search over all models can outperform the search of a FS-routine. One improves the fit to the data at hand, but this will not at all give the corresponding gain in the forecasting of new data. See Berk (1978).

3.7 Some alternative selection criteria

Model selection can be based on subjective judgement as well as on more objective methods. Often the two are combined. The objective methods for model selection have largely been based on either a testing approach or on a predictive approach. Our methodology belongs to the predictive side.

The prime examples of testing are F-tests between nested models in regression which is a specialization of likelihood ratio tests (Cox and Hinkley, 1974) between nested models in general. By nested we mean that one model is a reduction of the other.

The predictive tradition uses instead an estimate of performance for each model or method and selects the optimal. Akaike (1973) defined the most well-known criterion as $\text{AIC} = -\ln L + p$, where L is the likelihood for an estimated model with p parameters. For n normally distributed and independent observations on a linear model with constant variance, a monotone transform of AIC gives approximately

$$\text{AIC}' = s^2(1 + \frac{p}{n}) = \frac{\text{RSS}}{n}\frac{1 + p/n}{1 - p/n}$$

and this form has been widely used. Sometimes the approximation $(\text{RSS}/n)(1 + 2p/n)$ is used instead. We mentioned in Section 3.5 that for a correct regression model with estimated parameters the forward prediction error, which is another name for AIC', is an unbiased estimate of the expected mean square error when the model predicts new observations in the same design points. However, model selection with this measure is not consistent in the sense that it asymptotically finds the minimal correct model when such a model exists together with many overparametrized models in the model set. Such consistency can instead be achieved in regression and time series models by the measure $\text{BIC} = -2\ln L + p\ln n$. This form is derived from different points of view by Schwarz (1978), Akaike (1977), and Rissanen (1978). Cancelling uninteresting terms of $\exp(\text{BIC})$ and approximating the result slightly we have the approximate equivalent

$$\text{BIC}' = s^2(1 + \frac{p\ln n}{n}),$$

for standard multiple regression models.

Hannan and Quinn (1979) derived another consistent version, with a certain asymptotic optimality, on the form $\ln s^2 + \frac{cp\ln\ln n}{n}$, $c > 0$ and somewhat arbitrarily recommended c to be above 2.

Discussion

The AIC' criterion is exactly or approximately an unbiased measure for each correct model by itself. However, when we select between many models this measure is no longer unbiased for the *selected* model. Figure 3.3 shows what may happen with the measure in such a situation, and for finite n the BIC'-measure may fall into the same trap, as was demonstrated for time series by Hjorth (1982). However, the asymptotic consistency of BIC-measures is theoretically correct, but in the model selection field we have to be sceptical against asymptotic results when we are analysing data sets of say 50 or even 500 observations. The asymptotics may require several powers of ten more data before a reasonable accuracy is achieved, if many models are possible. As a practical rule of thumb, one can say that as long as the model selection affects the measure, so that $\text{CMV}(p)$ differs from $\hat{Q}_{\text{CV}}(p)$, such asymptotic results cannot be valid approximations. This can be contrasted with the central limit theorem, where a useful accuracy is often achieved by about ten observations. These observations were in fact the motivation behind our first attempts to use computer intensive methods for this problem. The same type of findings were recently reported by Breiman (1992) and Breiman and Spector (1992). The general ideas of their works are in fact very close to ours here and in Hjorth (1982, 1989), but they also extend the discussion. A simulation of the bootstrap type supports the model selection in Breiman (1992), and cross validation shows good performance in the extensive simulations of Breiman and Spector (1992).

In a more limited modelling, such as fitting autoregressive models of increasing order, the AIC' or BIC' can be quite useful since we then have only one model for each p and therefore get unbiased estimates s^2 for all the true and overfitted models.

3.8 A critical look at validation

The cross validation technique is relatively new and must be measured against more established statistical methods. Since it takes rather extensive computations it must provide something essential compared to the classical methods.

The elementary analysis in Example 3.1 was criticized earlier. The measure PRESS in Example 3.2 is usually rather close to $s^2(1 + p/n)$. The analysis in Copas (1983, Section 4) gives a related result and a similar asymptotic relation was derived by Stone (1977) between a kind of cross validation (without model variation) and Akaike's measure $AIC = \ln L - p$, where $\ln L$ is the logarithm of the maximal likelihood value for a model with p parameters. If we are estimating parameters by maximum likelihood, a simple adjustment of $\ln L$ will give about the same information as a cross validation (without model variation) in situations covered by Stone's analysis.

Validation with model variation, CMV, gives information which at present is not available by more analytical methods. Since it contains all the elements of estimation and selection it is more closely unbiased for the full procedure. The variance of the resulting CMV measure can be rather large sometimes, but is not given from the computations. At the end of Chapter 6 we give one method of estimating it by bootstrap methodology.

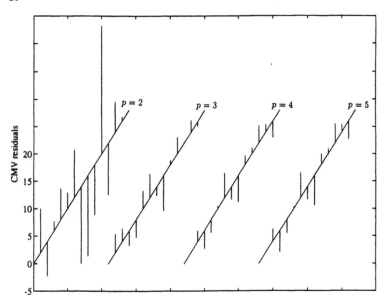

Figure 3.6 CMV *prediction errors for the Hald data and the model sizes $p = 2$ to $p = 5$.*

Finally, a practical argument for cross validation. The validation technique replaces theoretical computations by forecasts. Such forecasts are much easier to grasp for the nontheorist. The sequence of forecasts and errors which build up the CMV-measures can be written out or displayed graphically. Doing so they will give a direct illustration of the estimated precision and the error distribution. The qualitative difference between such errors and the residuals of traditional model fit must however be pointed out carefully. The traditional residuals are typically smaller and look better to the eye, but this is a false precision if model selection is involved with many models.

3.9 A meteorological data set

75	9							
35.714	2.060	5.427	20.	30.	30.	40.	45.	1.
71.841	0.879	4.240	7.	30.	70.	100.	100.	1.
81.111	0.882	4.283	5.	5.	30.	40.	45.	1.
122.619	0.803	4.501	4.	9.	90.	250.	300.	1.
128.333	1.416	5.874	1.	7.	60.	150.	250.	1.
283.333	0.814	4.857	1.	45.	50.	50.	150.	1.
42.897	0.865	4.071	7.	50.	60.	80.	100.	1.
19.167	1.921	6.119	20.	30.	50.	70.	90.	1.
6.667	3.169	6.043	20.	100.	200.	300.	300.	1.
13.667	1.473	4.029	8.	50.	70.	300.	500.	1.
11.389	2.092	5.653	20.	20.	20.	100.	200.	1.
203.733	6.547	8.739	1.	10.	100.	200.	250.	1.
57.000	5.996	9.534	5.	90.	150.	200.	200.	1.
244.444	0.730	4.992	3.	25.	30.	40.	25.	1.
7.750	1.738	6.717	20.	25.	30.	35.	40.	1.
195.833	0.815	4.947	2.	3.	25.	60.	120.	1.
29.286	3.108	7.676	20.	30.	80.	100.	100.	1.
65.750	0.950	3.946	8.	35.	100.	200.	260.	1.
57.250	0.814	4.812	2.	6.	50.	150.	200.	1.
76.667	2.111	5.826	10.	25.	60.	70.	120.	1.
240.000	0.891	4.413	1.	20.	100.	150.	200.	1.
11.567	1.733	6.959	18.	20.	30.	45.	60.	1.
145.833	0.727	5.265	8.	5.	5.	45.	80.	1.
31.500	0.724	5.356	5.	35.	60.	40.	50.	1.
51.444	0.637	5.585	6.	45.	200.	250.	400.	1.
27.300	0.620	5.860	4.	20.	200.	200.	200.	1.
73.333	3.995	12.180	6.	250.	300.	300.	400.	1.
120.000	0.531	6.136	3.	20.	50.	40.	40.	1.
47.000	1.252	8.303	20.	30.	40.	40.	40.	1.
80.333	0.728	5.219	3.	3.	3.	8.	40.	1.
51.900	5.373	10.530	2.	45.	100.	100.	200.	1.
48.833	5.364	10.621	3.	50.	80.	150.	200.	1.
137.333	0.728	5.219	3.	5.	60.	200.	200.	1.
22.500	1.273	6.797	6.	20.	100.	80.	80.	1.
187.333	0.633	5.677	2.	2.	2.	5.	15.	1.
20.157	3.971	12.271	6.	150.	150.	300.	300.	1.
72.879	1.232	8.426	10.	12.	20.	25.	40.	1.
29.524	1.730	7.020	20.	70.	100.	250.	250.	1.
100.967	0.635	5.631	1.	2.	35.	80.	300.	1.
35.700	4.696	11.353	3.	5.	150.	200.	300.	1.
261.111	0.630	5.723	1.	1.	3.	250.	500.	1.
71.917	0.620	5.860	3.	20.	60.	80.	80.	1.
66.667	0.617	5.906	3.	1.	3.	20.	100.	1.
107.333	0.613	5.952	2.	20.	250.	250.	250.	1.
9.250	1.294	7.997	20.	60.	100.	120.	120.	1.
49.083	0.514	6.319	3.	10.	20.	40.	100.	1.
157.500	0.499	6.457	1.	1.	1.	2.	3.	1.
225.556	5.380	10.439	18.	50.	100.	150.	200.	1.
17.317	4.570	11.904	6.	150.	400.	750.	750.	1.
17.533	4.040	11.996	6.	8.	300.	500.	750.	1.
102.778	0.509	6.365	1.	1.	200.	400.	400.	1.
21.095	1.715	7.263	15.	10.	20.	30.	35.	1.
42.262	0.637	5.585	3.	5.	5.	15.	60.	1.
51.133	4.679	11.445	2.	2.	150.	200.	300.	1.
156.191	0.541	5.998	1.	10.	3.	50.	60.	1.
73.500	5.373	10.530	8.	10.	150.	150.	200.	1.
25.466	0.728	5.219	6.	100.	500.	500.	500.	1.
15.079	1.514	7.508	20.	35.	40.	40.	30.	1.
44.444	1.509	7.569	20.	40.	60.	80.	100.	1.
106.667	4.040	11.996	1.	2.	2.	300.	300.	1.
52.000	3.995	12.180	9.	3.	2.	4.	25.	1.
56.250	3.971	12.271	5.	50.	200.	200.	200.	1.
20.000	1.733	6.959	20.	25.	20.	25.	30.	1.
34.133	5.341	10.804	3.	1.	6.	60.	120.	1.
168.667	0.523	6.228	6.	10.	5.	5.	60.	1.
32.787	1.753	5.356	6.	5.	40.	60.	70.	1.
42.000	3.834	12.730	4.	4.	4.	3.	35.	1.
94.000	1.253	9.428	15.	1.	6.	100.	80.	1.
65.833	0.921	10.244	20.	35.	40.	50.	60.	1.
13.500	0.710	9.876	20.	25.	25.	30.	25.	1.
16.000	1.273	11.133	15.	40.	80.	100.	100.	1.
34.333	0.797	9.398	15.	40.	100.	90.	100.	1.
139.216	0.315	7.184	3.	6.	25.	25.	20.	1.
227.000	0.285	7.451	1.	2.	6.	6.	2.	1.
68.583	0.218	7.713	2.	2.	2.	2.	8.	1.

3.10   CMV FORTRAN code

Validation of model size

```
C       Cross validation routine with the criterion function PRESS,
C       (replacable by the faster REG as indicated), and a model set
C       defined by a DESIGN matrix.
C       Presently limited to 11 variables including Y and
C       the constant (optional).
C
        DIMENSION D(250,11),XX(11,11),XXJ(11,11),B(11),CIN(512)
     1  ,CV(11),IDES(512,11),IPRED(11), ERROR(250,11)
        CHARACTER*30 DATFIL, DESFIL

        WRITE(*,*)' '
        WRITE(*,*)' '
        WRITE(*,9)
9       FORMAT(' KOVA.FOR IS A CROSS VALIDATION ROUTINE. ',/,
     1  ' IT READS A DATA FILE AND NEEDS ALSO A DESIGN FILE ',/,
     2  ' WHICH DEFINES THE SET OF POSSIBLE MODELS.',/,
     3  ' THE DATA FILE IS ORGANIZED IN THE FOLLOWING WAY: ',/,
     4  ' ROW 1 GIVES NR OF OBSERVATIONS, VARIABLES)',/,
     5  ' EACH ROW THEN HAS VAR1 VAR2,.. OF ONE OBS. ',/,
     6  ' THE DESIGN FILE HAS THE NR OF MODELS IN ROW ONE ',/,
     7  ' EACH MODEL HAS THE NR OF PREDICTORS ON ONE ROW',/,
     8  ' AND COLUMN NRS FOR THE PREDICTORS ON THE NEXT.',/,
     9  ' LIMITATIONS: 11 VARIABLES INCL. Y AND CONSTANT.',/,
     +  ' 250 DATA. (CAN BE ADJUSTED IN THE PROGRAM.')
        WRITE(*,*)' '
        WRITE(*,*)' GIVE THE NAME OF YOUR DATA FILE'
        READ(*,2)DATFIL
2       FORMAT(A)
        OPEN(22,FILE=DATFIL,STATUS='UNKNOWN')
        WRITE(*,*)' '
        WRITE(*,*)' GIVE THE NAME OF YOUR DESIGN FILE'
        READ(*,2)DESFIL
        OPEN(24,FILE=DESFIL,STATUS='UNKNOWN')
        READ(22,*)NOBS,NVAR
        FNOBS=FLOAT(NOBS)
        READ(22,*)((D(I,J),J=1,NVAR),I=1,NOBS)
        WRITE(*,*)'data'
        WRITE(*,*)NOBS,' DATA,   ',NVAR,' VARIABLES'
        WRITE(*,*)'DO YOU WANT DATA LISTED? 1=YES, 0=NO'
        READ(*,*)IMO
        IF(IMO.LE.0)GOTO 103
        DO 3 I=1,NOBS
3         WRITE(*,*)(D(I,J),J=1,NVAR)
103       CONTINUE
        READ(24,*)NMOD
        DO 5 N=1,NMOD
        READ(24,*) IDES(N,1)
        JJ=IDES(N,1)
5         READ(24,*)(IDES(N,J),J=2,JJ+1)
        WRITE(*,*)'WHICH COLUMN NUMBER IS Y?'
        READ(*,*)IY

        DO 10 I=1,NVAR
        DO 10 J=1,I
        DO 20 K=1,NOBS
20        XX(I,J)=XX(I,J)+D(K,I)*D(K,J)
10        XX(J,I)=XX(I,J)
```

```
          DO 25 IP=1,NVAR-1
    25       CV(IP)=0.

          WRITE(*,*)' LOOP OVER ALL DATA'
          DO 300 JUT=1,NOBS
          WRITE(*,*)JUT
          DO 30 I=1,NVAR
          DO 30 J=1,NVAR
    30       XXJ(I,J)=XX(I,J)-D(JUT,I)*D(JUT,J)
    C     LOOP OVER MODEL SIZES
          DO 250 IP=1,NVAR-1
          COPT=.9999E32
    C     LOOP OVER MODELS OF SIZE IP
          DO 200 N=1,NMOD
          IF(IDES(N,1).NE.IP)GOTO 200
    C     **********************************************
    C     **** THE FUNCTION PRESS CAN BE REPLACED BY E G
    C     **** REG(XXJ,IY,IPRED,NP,B,CRIT)
          DO 40 I=1,IDES(N,1)
    40       IPRED(I)=IDES(N,I+1)
          NP=IDES(N,1)
          JT=JUT
          CALL PRESS(D,NOBS,NVAR,XXJ,JT,IY,IPRED,NP,CRIT)
    C     **********************************************
          CIN(N)=CRIT
    C     FIND THE BEST MODEL OF SIZE IP
          IF(CIN(N).GT.COPT)GOTO 200
          COPT=CIN(N)
          NOPT=N
    200      CONTINUE
    C
    C     NOPT IS FOUND TO BE THE BEST MODEL OF SIZE IP
          NP=IDES(NOPT,1)
          DO 70 I=1,NP
    70       IPRED(I)=IDES(NOPT,I+1)
          CALL REG(XXJ,IY,IPRED,NP,B,VAR)
          Y=D(JUT,IY)
          E=Y
          DO 80 I=1,NP
    80       E=E-B(I)*D(JUT,IPRED(I))
          ERROR(JUT,IP)=E
          CV(IP)=CV(IP)+E**2/FNOBS
    250      CONTINUE
    300      CONTINUE
    C
    302      WRITE(*,*)'ARE YOU READY FOR RESULTS? 1=YES'
          READ(*,*)IMO
          IF(IMO.LE.0)GOTO 302
          WRITE(*,*)' '
          WRITE(*,*)'MODEL SIZES AND CMV(IP)'
          DO 310 I=1,NVAR-1
    310      WRITE(*,*) I,CV(I)
          CVOPT=.9999E32
          DO 320 I=1,NVAR-1
          IF(CV(I).GT.CVOPT)GOTO 320
          CVOPT=CV(I)
          IPOPT=I
    320      CONTINUE
          WRITE(*,*)' '
          WRITE(*,*)' OPTIMAL MODEL SIZE =    ',IPOPT
          WRITE(*,*)' '
          WRITE(*,*)' BEST VALIDATION MEASURE=',CVOPT
          WRITE(*,*)' '
```

```
          WRITE(*,*)'DO YOU WANT THE MODEL? 1=YES, 0=NO'
          READ(*,*)IMO
          IF(IMO.LE.0)GOTO 435
   C      FINAL SELECTION OF MODEL OF SIZE IPOPT (ALL DATA)
          NP=IPOPT
          COPT=.9999E32
          DO 400 N=1,NMOD
          IF(IDES(N,1).NE.IPOPT)GOTO 400
          DO 410 I=1,NP
   410        IPRED(I)=IDES(N,I+1)
   C      ****************************************************
   C      REPLACEABLE BY E G  REG(XX,IY,IPRED,NP,B,CRIT)
          CALL PRESS(D,NOBS,NVAR,XX,0,IY,IPRED,NP,CRIT)
   C      ****************************************************
          IF(CRIT.GE.COPT)GOTO 400
          COPT=CRIT
          NOPT=N
   400        CONTINUE
          DO 420 I=1,NP
   420        IPRED(I)=IDES(NOPT,I+1)
          CALL REG(XX,IY,IPRED,NP,B,VAR)
   C      OUTPUT OF FINAL MODEL
          WRITE(*,*)' '
          WRITE(*,*)'SELECTED MODEL, VARIABLE NR, COEFFICIENTS'
          DO 430 I=1,NP
   430        WRITE(*,*)IPRED(I),B(I)
          WRITE(*,*)' '
          S2KONV=VAR/(FNOBS-FLOAT(NP))
          WRITE(*,*)'s2 BY CONVENTIONAL REGRESSION ',S2KONV
   435        CONTINUE
          WRITE(*,*)' '
          WRITE(*,*)' DO YOU WANT CMV RESIDUALS? 1=YES,0=NO'
          READ(*,*)TEST
          IF(TEST.LE.0.) GOTO 440
          WRITE(*,*)' MODEL SIZES BETWEEN M1 OCH M2. GIVE M1,M2.'
          READ(*,*)M1,M2
          WRITE(*,*)' '
          WRITE(*,*)' '
          WRITE(*,*)' RESIDUALS FROM VALIDATION CROSS MODELS'
          DO 450 J=1,NOBS
          WRITE(*,8) J,(ERROR(J,M),M=M1,M2)
   450        CONTINUE
   8      FORMAT(I4,6F8.2)
   440        CONTINUE
          WRITE(*,*)' '
          WRITE(*,*)' '
          END
   C
          SUBROUTINE REG(R,IY,IPRED,NP,BETA,VAR)
          DIMENSION R(11,11),IPRED(11),BETA(11),A(11,11)
          DIMENSION B(11,11)
          DO 200 I=1,NP
          I1=IPRED(I)
          DO 200 J=1,NP
          J1=IPRED(J)
   200        A(I,J)=R(I1,J1)
          CALL INVERS(A,B,NP)
          DO 300 I=1,NP
          BETA(I)=0.
          DO 300 J=1,NP
   300        BETA(I)=BETA(I)+B(I,J)*R(IPRED(J),IY)
          VAR=R(IY,IY)
```

```
      DO 400 I=1,NP
400      VAR=VAR-BETA(I)*R(IPRED(I),IY)
      RETURN
      END
C
C
      SUBROUTINE INVERS(A,B,N)
      DIMENSION A(11,11),B(11,11),AB(11,22)
      DO 20 I=1,N
      DO 10 J=1,N
      AB(I,J)=A(I,J)
10       AB(I,N+J)=0.
20       AB(I,I+N)=1.
      DO 200 I=1,N
      C=SIGN(1.,AB(I,I))
      N2=N+N
      DO 80 L=1,N2
80       AB(I,L)=AB(I,L)/C
      IF(I.EQ.N) GOTO 50
      DO 100 K=I+1,N
      C=-SIGN(1.,AB(K,I))
      DO 90 L=1,N2
90       AB(I,L)=AB(I,L)-C*AB(K,L)
100      CONTINUE
50       CONTINUE
      C=AB(I,I)
      IF(ABS(C).LT..1E-30)
1        WRITE(*,*)((A(II,JJ),JJ=1,5),II=1,5),'N=',N
      IF(ABS(C).LT..1E-30)STOP
      DO 120 L=1,N2
120      AB(I,L)=AB(I,L)/C
      DO 150 K=1,N
      IF(K.EQ.I) GOTO 150
      C=AB(K,I)
      DO 140 L=1,N2
140      AB(K,L)=AB(K,L)-C*AB(I,L)
150      CONTINUE
200      CONTINUE
      DO 300 I=1,N
      DO 300 J=1,N
300      B(I,J)=AB(I,N+J)
      RETURN
      END
C
      SUBROUTINE PRESS(D,NOBS,NVAR,XX,JUT,IY,IPRED,NP,CRIT)
C     LET JUT=0 IF NO DATA WAS REMOVED EARLIER
      DIMENSION D(250,11),XX(11,11),IPRED(11),XXK(11,11),B(11)
      CRIT=0.
      FNOBS=FLOAT(NOBS)
      IF(1.LE.JUT.AND.JUT.LE.NOBS) FNOBS=FNOBS-1.
      DO 150 IUT=1,NOBS
      IF(IUT.EQ.JUT)GOTO 150
      DO 50 I=1,NVAR
      DO 50 J=1,NVAR
50       XXK(I,J)=XX(I,J)-D(IUT,I)*D(IUT,J)
      CALL REG(XXK,IY,IPRED,NP,B,VAR)
      Y=D(IUT,IY)
      E=Y
      DO 60 I=1,NP
60       E=E-B(I)*D(IUT,IPRED(I))
      CRIT=CRIT+E**2/FNOBS
150      CONTINUE
      RETURN
      END
```

3.11 Exercises

Exercise 3.1 A weather forecaster gives rain probabilities
in ten classes characterized by probabilities .05, .15, ..., .95.
A farmer stores data as $(y, \hat{p})$ where $y = 1$ if rain, 0 if dry,
and $\hat{p}$ is the forecast. His data are:

   (1, 0.55)(0, 0.15)(0, 0.05)(0, 0.45)(1, 0.45)(1, 0.75)
   (1, 0.25)(0, 0.25)(0, 0.05)(0, 0.05)(1, 0.35)(0, 0.35)

He intends to go on collecting data and improve the forecasts
by a model $\hat{p}_1 = \beta_0 + \beta_1 \hat{p}$. How should he estimate the
parameters and validate whether his result seems to be better
than the given forecast?

Exercise 3.2   Discuss different ways to shrink estimated
parameters in the regression model $Y = \beta_0 + \beta_1 x_1 + \ldots + \beta_m x_m + \varepsilon$
   Could you use different procedures when variables are en-
tered one at a time as in stepwise regression compared to
when all variables are entered directly? How do you validate?

Exercise 3.3 What properties in the data or the model will
cause a strong (or weak) shrinking factor? No computations;
use your intuition and guess.

Exercise 3.4 Use CMV to select a prediction model for column 1, a derived visibility variable, for the meteorological
data set given before. Neglect the possible time dependency
of the data. The exercise can be varied by for example allowing preceding values of column 1 as extra predictors. This
only costs a few data rows and some editing. If the listed
CMV-program is used, a special file DESIGN.DAT has to be
created to describe all candidate models. In this file, line 1
gives the number of models. Then each model gets two lines,
one for the number of predictors, and the other for the column numbers of the predictors. The program will ask for the
column number of the Y-variable. The data should be given
in free format, one observation per row. Before all data, one

Table 3.6 *A bivariate data set.*

x	y	x	y	x	y	x	y
3.81	6.2	4.50	6.4	4.96	5.2	4.34	6.5
4.88	4.9	3.80	6.3	5.04	5.0	2.25	6.2
4.59	6.5	5.63	4.0	4.56	6.1	3.19	6.1
4.52	5.7	3.12	5.6	4.26	5.6	2.83	6.0
2.75	6.1	4.88	5.2	4.46	6.1	4.68	5.6
4.81	5.2	0.91	5.6	2.85	6.5	4.69	5.8
3.91	6.4	4.04	6.4	2.50	5.8	3.24	6.0
4.34	6.5	3.63	6.7	1.92	5.4	3.15	7.1
5.72	3.8	4.55	6.0	4.75	5.2	4.53	5.8
3.21	6.3	5.28	5.1	4.04	6.4	3.90	6.1
4.32	5.8	5.46	4.5	2.45	5.5	4.11	6.3
5.72	3.7	3.67	5.7	3.96	7.3	4.40	6.2
4.47	5.9	5.34	4.3	3.14	6.0	3.28	6.0
4.43	6.6	3.12	6.1	5.02	5.3	2.39	6.1
4.86	5.5	3.89	6.7	3.51	6.4	3.69	6.3
3.55	6.2	3.61	6.5	5.69	4.1	3.45	6.1
3.84	6.2	3.07	6.0	3.39	6.3	5.09	5.3
3.29	6.1	5.14	5.7	3.05	5.9	4.80	6.0
4.56	5.5	2.76	6.0	5.25	5.6	2.13	6.2
3.73	6.3	4.57	5.8	4.67	5.8	3.11	6.5
5.33	5.1						

row should give the number of data items and the number of columns.

Exercise 3.5

In Table 3.6, a sample (x_i, y_i), $i = 1, \ldots, n$, of independent bivariate data has a non-standard distribution as can be seen from a plot. We can estimate the probability density function by a kernel method

$$\hat{f}(x, y) = \frac{1}{n} \sum_{i=1}^{n} k(x - x_i, y - y_i; \theta),$$

where k can be chosen as a Gaussian kernel, $k(u, v; \theta) = \frac{1}{2\pi\theta^2} \exp(-(u^2 + v^2)/2\theta^2)$, with parameter θ. (If x and y are on very different scales, a generalization with two parameters θ_1, θ_2 can be preferred). Delete one data point at a time and compute $\hat{f}$ on the rest. Find a way to evaluate this estimate on the excluded data, i.e. invent a loss or gain function. In case you find no such possibility after some thought, use the appendix of Chapter 4 and see if that helps. Perform a full cross validation. Try some different values of θ and compare. You are now on the road to the estimation of an optimal θ.

CHAPTER 4

Validation of time series problems

4.1 Introduction

A series of successive and regularly taken measurements is called a time series. Typical examples from the field of economics are quarterly sales figures, daily or weekly foreign exchanges, export volumes and various other financial indices. The collection and analysis of such series plays an important role in the work of many decision makers. The large scale developments are often used as a basis for decisions about investments or cuts of production, and on a shorter time scale they assist in decisions on buying and selling. In industry, time series may have even more frequent use for routine decisions about the planning and control of production and processes. There are many other areas where time series are extensively used. The vast amount of data of a physical nature, collected in meteorology and other earth sciences, is worth mentioning.

Analyses of time series are made at very different levels of sophistication. A director may look at some drawn series and make his purely subjective analysis by eye. This represents one extreme. At the other end of the spectrum an analysis group may fit sophisticated stochastic models and optimize in terms of these. There are some good reasons for fitting models. One is that forecasts and other analyses become objective and unaffected by optimistic or pessimistic moods. A second reason can be that decisions are made so often that a subjective technique becomes cumbersome and/or will be less carefully made. A third and more important reason can be that the models give better forecasts and will find relations which are not easily observed bye eye. However, there are also situations where a subjective method has an advantage, e.g. by considering information of an occasional nature.

Today several program packages are available which can handle time series by more or less general classes of models. The prime example is the popular class of 'autoregressive integrated moving average' (ARIMA) models and the sub-classes ARMA, AR-, and MA-models. On the other hand there is a shortage of programs and methods which evaluate correctly the analyses made. (Hopefully we will contribute something here.) Often the programs will only show fits to the same data as the models were estimated on, and this will typically give a too optimistic picture of the results.

Any serious user of standard packages must make the effort to get a clear picture of the class of models used and these models' limitations. Also the way uncertainties are produced (if produced at all) is very important, but it takes some knowledge to understand the different possible versions.

In this chapter we will first present a couple of different models in order to make the following discussion more concrete. We then show how the cross validation technique is fairly easily modified into forward validation of time series forecasts. We develop general methods for model selection and comparison of model classes based on such validation and give some examples. This is the main message of this chapter. Finally we give some different loss functions for various kinds of forecasts, the reason being that the imagination should not be limited to just scalar forecasts and squared errors. The loss functions are applicable also to Chapter 3.

4.1.1 Terminology

A time series $\dots y_1, y_2, \dots, y_t, \dots$ is *scalar* valued if just one value is studied each time, otherwise the series is *vector valued* or *multivariate*.

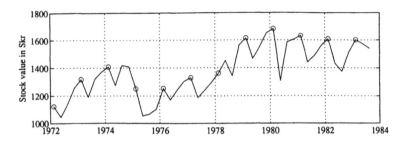

Figure 4.1 *Quarterly data on Swedish paper and board production*

A model $\ldots Y_1, Y_2, \ldots, Y_t, \ldots$ for the series is *stationary* if its probability distributions do not change when the time parameter is translated. Otherwise the model is non-stationary.

If we regard models for economic time series, the models will usually have to consider both *seasonal variation* and *trend*. By trend we mean a (relatively) systematic increase or decrease with time. Such effects can also be present in physical and technical series. At best one can then fit a stationary model after correction for such non-stationary effects. The most interesting feature of time series models is however their modelling of the dependency between Y_t for different t.

4.2 Model examples

In Figure 4.1 we show quarterly values of Swedish paper and board production. We are looking for a good model predicting the production volume one time step ahead. The only information used in the forecast will be the series itself. (More realistic multivariate situations are saved for later examples.)

In the figure we can observe that the values in quarter II are systematically low and that the process has largely been growing during 1972 to 1983. This motivates a scalar model with seasonal effects and positive (increasing) trend as one possible structure for the series.

4.2.1 Scalar model, type one

This model has additive deterministic trend and seasonal and autoregressive random variation.

$$Y_t = a + bt + S_{k(t)} + X_t$$
$$X_t = c_1 X_{t-1} + \ldots + c_r X_{t-r} + \varepsilon_t \tag{4.1}$$

$t = 1, 2, 3, \ldots$	time enumeration
$k(t) = \{1, 2, 3, 4\}$	quarter at time t
$S_1 + S_2 + S_3 + S_4 = 0$	seasonal effects adding to zero; four seasons in this example
$a + bt$	linear trend
$c_1, \ldots, c_r$	constants with a side condition allowing X_t to be stationary
ε_t	independent identically distributed deviations (often assumed $N(0, \sigma)$)

For the mathematically interested reader, we can mention that a stationary X_t-process exists if and only if all roots $z_1, \ldots, z_r$ to the equation $1 - \sum_1^r c_n z^n = 0$ have $|z_k| > 1$, i.e. lie outside the unit circle in the complex plane. With $r = 1$ this means that $X_t = c X_{t-1} + \varepsilon_t$ where $|c| < 1$.

We can eliminate X_t in the model and write it more compactly as

$$Y_t - S_{k(t)} = \sum_{i=1}^{r} c_i (Y_{t-i} - S_{k(t-i)}) + A + Bt + \varepsilon_t \tag{4.2}$$

with new parameters A and B.

A direct analysis is somewhat complicated by the product $c_i S_{k(t-i)}$, at least if we want to use certain standard programs for the analysis. Traditionally, one usually avoids this problem by first subtracting out estimated trend and seasonal effects. After this subtraction an autoregressive model is fitted to the rest of the series. The trend and seasonal effects can be estimated by regression using the upper model line in (4.1). At this estimation one neglects the dependency between different time points. The residuals of this analysis are treated as X_t and modelled according to the second line of (4.1). Also in this step the parameters c_i can be estimated by regression analysis (least squares estimates). This is accomplished by solving the least squares estimates $\hat{c}_1, \ldots, \hat{c}_r$ from the expression

$$\begin{pmatrix} X_{r+1} \\ X_{r+2} \\ \vdots \\ X_n \end{pmatrix} = \begin{pmatrix} X_r & X_{r-1} & \cdots & X_1 \\ X_{r+1} & X_{r-2} & \cdots & X_2 \\ \vdots & \vdots & & \vdots \\ X_{n-1} & X_{n-2} & \cdots & X_{n-r} \end{pmatrix} \begin{pmatrix} c_1 \\ c_2 \\ \vdots \\ c_r \end{pmatrix} + \begin{pmatrix} \varepsilon_{r+1} \\ \varepsilon_{r+2} \\ \vdots \\ \varepsilon_n \end{pmatrix}$$

If the series is short, somewhat better estimates can usually be achieved by the maximum likelihood method but we will not go into this here. From the regression analysis only the estimates should be used. Estimated uncertainties from a traditional regression model can be completely wrong due to dependency between the Y_t and should be neglected. Confidence regions for c-parameters are discussed in Jenkins and Watts (1968) and give valid estimates when a single given model is estimated. In the model selection context, everything changes and no classical estimates of uncertainty can be trusted.

Some variations of the model are possible. The expression $a + bt$ for the trend can be replaced by the constant only, or by higher degree polynomials, e.g. $a + bt + ct^2$. Also cyclic trends like $a + bt + d \cos(\lambda(t - t_0))$ have been tried on economical series to my knowledge with little predictive success.

4.2.2 Multiplicative effects

If effects and variations are better measured as proportions of the series than as additive effects, we can modify the first type of model accordingly. Instead of additive effects as in model (4.1) we can model multiplicative effects, and for example write the upper model line as

$$Y_t = e^{a+bt+S_{k(t)}} + X_t. \qquad (4.3a)$$

If Y_t is always positive we can replace X_t by $\exp(X_t)$ and analyse the computationally simpler structure of the logarithms

$$\ln Y_t = a + bt + S_{k(t)} + X_t. \qquad (4.3b)$$

In both cases we may use the autoregressive model $X_t = c_1 X_{t-1} + \ldots + c_r X_{t-r} + \varepsilon_t$ as in (4.1) for the random variation. At the cost of more complicated estimation some other ARIMA model could also be used for the X_t.

4.2.3 Scalar model, type two

In practice, models with deterministic trends and seasons cannot be expected to remain valid in for example the economy during a very long time. Cyclical models may be even worse in this respect. It is extremely difficult to decide which data are useful and which are too old for efficient estimation of a prediction model for the (nearest) future. If only very recent data are used, the estimation uncertainty becomes large, and with many old data we may come out with a model which is not up to date. It is possible (e.g. by validation) to optimize the age of the oldest data, but a more interesting alternative is to build adaptivity into the model.

The second kind of model we consider here has stochastic trend and stochastic seasonal effects. These quantities still exist in the model, but are redefined so that they may change with time and fit new situations. Independent noise is added. In this model type 'everything' is random, only the parameters of some probability distributions are stable.

Let μ_t be the 'level' of the process at time t, β_t the 'slope of the trend' at t, S_t the seasonal effect at t, and ε_t the noise.

Suppose we have quarterly data with four seasons (use 12 if monthly and 7 if daily seasonals are required). Write the model

$$Y_t = \mu_t + S_t + \varepsilon_t \qquad \varepsilon_t \text{ independent } N(0, \sigma_1)$$
$$\mu_t = \mu_{t-1} + \beta_{t-1} + \nu_t \qquad \nu_t \text{ independent } N(0, \sigma_2)$$
$$\beta_t = \beta_{t-1} + \xi_t \qquad \xi_t \text{ independent } N(0, \sigma_3)$$
$$S_t = -S_{t-1} - S_{t-2} - S_{t-3} + \eta_t \qquad \eta_t \text{ independent } N(0, \sigma_4).$$
$$(4.4)$$

This model was suggested by Harvey (1989). If $\sigma_4 = 0$ the seasonal effects will be constant, if $\sigma_3 = 0$ the slope β of the trend becomes constant, and if $\sigma_2 = 0$ the level is constant. Deterministic effects are therefore a special case of this model (at the boundary of the parameter space). The only parameters of this model are $\sigma_1, \sigma_2, \sigma_3$ and σ_4. In principle an autoregressive structure for ε_t can be assumed at the cost of some extra parameters. The quantities μ_t, β_t, S_t are stochastic and may be estimated by Kalman filter techniques. Such estimates can be made recursively, working through the data in the given order, or on line as new data is entered. The recursive method is well suited to the forward validation method of this chapter. Starting values or assumed initial normal distributions or the equivalent are required for $\mu_0, \beta_0, S_0, S_{-1}, S_{-2}$ (four seasons) and will act approximately as extra parameters.

Several other univariate time series models can be defined. Some of these are easily available in textbooks; see the end of the next section. Here we only give a minimal background for the study of model selection and validation techniques. Moreover the models given here are flexible enough for a wide class of applications. The most important extension in practice is given in the next section.

4.2.4   A multivariate model

When several time series are studied simultaneously we have a multivariate series and the number of conceivable models increases dramatically. It is still a fairly simple task to search prediction models of the regression type with deterministic seasonal effects and trends. Each series may for example be

seasonally adjusted and trend adjusted on its own. The need
for validation techniques is, however, especially pronounced
at multivariate series due to the large model set and the
difficulty of selecting a proper size for the model.

The autoregressive model of type (4.2) can be generalized
to the following vector model

$$\mathbf{Y}_t - \mathbf{S}_{k(t)} = \sum_{i=1}^{r} \mathbf{C}_i(\mathbf{Y}_{t-i} - \mathbf{S}_{k(t-i)}) + \mathbf{A} + \mathbf{B}t + \varepsilon_t$$

where $\mathbf{Y}., \mathbf{S}., \mathbf{A}, \mathbf{B}, \varepsilon$ are vectors, $\mathbf{C}_i$, $1 \le i \le r$, are square
matrices and ε_t is often assumed to be independent $N(\mathbf{0}, \mathbf{D})$
where $\mathbf{D}$ is a covariance matrix. The number of parame-
ters is considerable since the matrices $\mathbf{C}_i, \mathbf{D}$ and the vectors
$\mathbf{A}, \mathbf{B}, \mathbf{S}_k$ have to be estimated. Often one therefore searches
thinned prediction formulas for one $\mathbf{Y}_t$-component at a time
such as, for component r,

$$\hat{Y}_{r,t} = S_{r,k(t)} + \sum_{j=1}^{p} e_j(Y_{r_j,t-i_j} - S_{r_j,k(t-i_j)}) + A + Bt. \quad (4.5)$$

The index r_j denotes a selection of one component of the Y-
vector and $t-i_j$ a selection of one of the time-lags considered.
The estimation and subtraction of trend and season is based
on all the observations, but at the validation they must be
computed anew for shorter sections of the time series. We
may of course let multiplicative seasonal effects and trends
replace the additive effects also for multivariate series. The
difference in computational effort becomes minimal.

The reader interested in a more extensive discussion on
different time series models is referred to specialist books on
the subject, such as Chatfield (1980), Kendall (1976), and
Hjorth (1987, in Swedish). The class of ARMA or ARIMA
processes has become very popular, mainly due to the book
by Box and Jenkins (1970). The time series models are part

of the theory of stochastic processes, which is a very active area in probability and statistics with a rich literature in books and journals.

4.3 Forward validation

Time series models are built primarily for two reasons. One is prediction of later values, which has obvious strategic and economic interest. The other is description in a theoretical manner of the historical development in order to try to learn something from the description. The ability to predict is of course most important if the purpose is to make predictions. In the description of the history one often aims at catching the main features of the development with as few parameters as possible. In order to make an objective comparison between alternative descriptive models one often uses predictability as a criterion also in this case. The idea is that a good description or interpretation will automatically give good forecasts. On the contrary, models with good predictive properties are not always easy to interpret.

In model selection based on validation it is the predictability only we are studying. All other properties of interest have to be considered outside the validation. We can do this either by only allowing acceptable models to compete against each other or by post checking, where preliminary selected models can be taken out due to defects which do not disturb the prediction measures.

A time series $y_1, y_2, \ldots, y_n$ is observed. The series may be scalar or multivariate, but we use the scalar notation here with lower case letters for observed values. In the validation of any time series models it is natural to regard the dependency and therefore follow the natural direction of time. We then let the validation build upon real forecasts. Introduce the symbols

$ES_t = \{y_1, \ldots, y_{t-1}\}$ observations up to $t - 1$,

$\{\alpha, \quad \alpha \in A\}$ the set of alternatives validated,

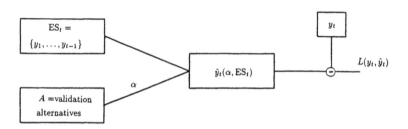

Figure 4.2 *Computations after time* $t-1$

$\hat{y}_t(\alpha, \mathrm{ES}_t)$ predictions of y_t for the alternative
                                  α estimated on ES_t,
$L(y_t, \hat{y}_t)$ loss function for prediction errors.

The symbol ES_t denotes the set of observations available for
a (one step) prediction of y_t. This means that ES_t corresponds to the ES_t of Chapter 3, and y_t corresponds to TS_t
there, but the analogy is not perfect since $\mathrm{ES}_t \cup \mathrm{TS}_t = \mathrm{ES}_{t+1}$
here whereas the same union gives the whole data set DS in
Chapter 3. The validated alternatives α in the set A can be
of a different nature due to the problem. They can be quite
different models, or different model sizes at a stepwise search
of models, or the value of a particular parameter in a certain kind of model. (We need a restriction on A later due to
our definition of the weights. Each α should correspond to a
number $p = p(\alpha)$ of parameters. This restriction is not fundamental since the validation could be somewhat modified
if this is not fulfilled.) If there is just one alternative in A,
the validation is reduced to an evaluation of the predictability of that alternative. Otherwise it evaluates the process of
both estimating models and selecting one alternative. The
validation works as follows.

After time $t - 1$, all models corresponding to alternatives
$\{\alpha, \alpha \in A\}$ are estimated on ES_t. Their predictions are
then tried on the next observation y_t. This is schematically
illustrated in Figure 4.2.

4.3.1 Selection procedure

We need some data before all models can be estimated. Let $m - 1$ observations be sufficient and start the validation on data item y_m. Define the *decision measure*

$$C(\alpha) = \sum_{t=m}^{n} \gamma_{t,n} L(y_t, \hat{y}_t(\alpha, \mathrm{ES}_t)) \qquad (4.6)$$

as a weighted average of the losses for each $\alpha \in A$. The weights $\gamma_{t,n}$ are given below. Finally, select the alternative $\alpha_0 = \alpha_0(\mathrm{ES}_{n+1})$ minimizing $C(\alpha)$. This defines the selection procedure.

4.3.2 Estimate of performance

The kind of selection made after time n could also be made at an earlier time such as just after $t - 1$. The measure (4.6) is then replaced by a shorter sum up to $n' = t - 1$. Let $\alpha_0(\mathrm{ES}_t)$ denote the result of this selection. Since $\alpha_0(\mathrm{ES}_t)$ can vary with t, statistical effects of model selection can now enter into the evaluation.

As an estimate of the selected model's expected loss another weighted sum is computed

$$\mathrm{CMF} = \sum_{t=m}^{n} \delta_t L(y_t, \hat{y}_t(\alpha_0(\mathrm{ES}_t), \mathrm{ES}_t)). \qquad (4.7)$$

(CMF = Cross Model Forward validation.)

This is our estimated performance measure for the selected alternative including the selection procedure. The difference

$$\Delta = \mathrm{CMF} - C(\alpha_0)$$

can be used as an estimate of the bias in the minimum of $C(\alpha)$. This bias depends on both the minimization of $C(\alpha)$ itself, the use of constant α through the time series and on the different weights used.

Under the assumption that the same bias $E[\Delta]$ is approximately valid for all α such that $C(\alpha)$ is close to $C(\alpha_0)$, i. e.

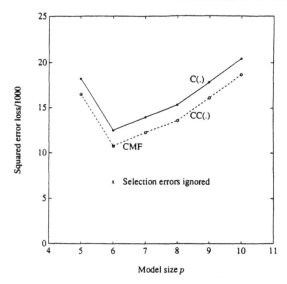

Figure 4.3 *Measures* C(α), CC(α) *and* CMF *of a forward validation on the paper and board data of Section 4.4.1*

for all α competing with α_0 after the analysis, we may prefer to display the measure

$$CC(\alpha) = C(\alpha) + \triangle, \quad \alpha \in A \qquad (4.8)$$

as an estimated performance measure for the different alternatives α in A. The bias correction is typically not relevant for large $C(\alpha)$ but this is of minor importance. The lack of information about where to stop the adjustment makes it better to use it either everywhere or not at all outside α_0. An example of the measures (4.6)–(4.8) is drawn in Figure 4.3.

The flow of computations is illustrated in a flow chart later in this chapter, and an example of a computer output is also given there.

4.3.3   Weights

The weights $\gamma_{t,n}$ used in (4.6) for model selection are normalized so they sum to 1 and are defined as

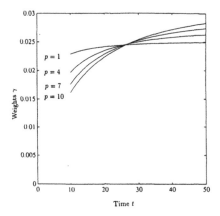

Figure 4.4 *Model selection weights γ for some model sizes p at times between $m = 10, n = 50$*

$$\gamma_{t,n} = \frac{1}{1 + \frac{p}{t-1}} \bigg/ \sum_{i=m}^{n} \frac{1}{1 + \frac{p}{i-1}} \qquad (4.9)$$

for $m \le t \le n$, see Hjorth (1982). Here we use the restriction that each validation alternative α in (4.6) corresponds to a given number $p = p(\alpha)$, describing the number of parameters used for the model(s) associated with α. The weights are illustrated in Figure 4.4.

The measure CMF in (4.7) is intended to give an estimate of the selected model's future performance. Here a second set of weights is used. A complication is that the alternatives $\alpha_0(ES_t)$ that appear to be optimal at time t can vary with t, and so we have no given constant p in the weights. Also the weights should sum to less than 1 since we have more information at the end of the time series than during the validation process. In Hjorth (1982) the following weights were used and motivated. The weights were first derived for a recursive solution of the regression problem and have been simulated in time series analyses. Let $p(t) = p(\alpha_0(ES_t))$ be the estimated optimal model size at time t and compute

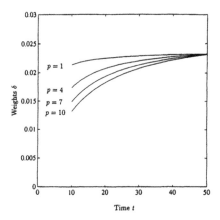

Figure 4.5 CMF *weights δ for some model sizes p at times between*
$m = 10$, $n = 50$

$$\delta_t = \frac{1 + p(t)/n}{(n - m + 3)(1 + \frac{p(t)}{t-1})}, \quad m \le t \le n. \qquad (4.10)$$

A sample of such weights is drawn in Figure 4.5. If $p(t)$
varies, the weights will jump between different curves of the
kind drawn.

In both cases the weights γ and δ are for a given p increasing with t. The intuitive reason is that the information in
ES_t grows and approaches the final information from the entire ES_{n+1}. Forecasting errors at the end are therefore more
relevant than earlier errors. Besides, this effect is stronger
the larger the number p of parameters we have to estimate.
The increased risk of large errors early on also explains why
the weights δ_t sum to less than 1 (for each fixed p). We are
in fact trying to estimate future performance of the selected
alternative.

A nice property of the weights is that the sums $C(\alpha)$ and
CMF (4.6)–(4.7) can be computed recursively. Consequently
there is no need for extra computation when these sums are
needed at intermediate times.

4.3.4 Forward validation flow chart

TIME LOOP

$t = m, \ldots, n$

LOOP OVER ALTERNATIVES α_i
$i = 1, \ldots, imax$

UPDATING MODEL ESTIMATES FOR α_i
FORECAST $\hat{y}_t = \hat{y}_t(\alpha_i, ES_t)$

LOSS $L(\hat{y}_t, y_t)$

UPDATING VALIDATION MEASURE

$p = p(\alpha)$ number of parameters

$g = \frac{1}{1 + \frac{p}{t-1}}$

$d_i = d_i + g$ normalizing sum

$V_i = V_i + gL(\hat{y}_t, y_t)$

$C_i = V_i/d_i$

α_{i_0} optimal alternative from preceding time

IF $i = i_0$ THEN CMF $=$ CMF $+ gL(\hat{y}, y)\frac{1+p/n}{n-m+3}$
end of i-loop

FIND NEW BEST i_0 MINIMIZING C_i

$\alpha_{i_0} = \alpha_{i_0}(ES_{t+1})$

end of t-loop

OUTPUT α_{i_0}, estimated model, CMF

4.3.5 Computer output example

We illustrate a forward selection analysis on some data presented later in this chapter, the paper and board data set. The program run is called FORVAL.FOR, and is coded in FORTRAN.

Looking at the results of the analysis, one notices in particular the very large model selection effects on the least squares measures. Compared to the prediction errors building up CMF and $C(\alpha)$, the least squares values are severe underestimates. This carries over to the Akaike measures which suggest too many parameters in the selected model.

Forward validation of time series.
Limitations: max 249 series
max 34 series
max 14 forced variables
quadratic or linear trend
max 199 predictor candidates
max 20 selected predictors
Row one of the data file must have
the following five numbers
nr of series (columns), nr of time points
(rows below the first), nr of seasons,
year (block), and season of first data
Give the name of your data file paper.dat
The file has
28 series observed at
47 time points. Times are divided into
4 seasons.
first year (block) and season 72 1
Which series (column) is predicted?   7
How many time-steps ahead?   1
How many series are used as predictors? 9
Give numbers of the predictor series   1 3 5 6 7 8 10 12 20
How many different predictor time lags? 4
Give the predictor lags, 0=present time 0 1 2 4
Quadratic trend? 1=yes, 0=no   0
Season and trend are forced into models
in all 41 predictors. 5 are forced
minimum 5 predictors in models.
Give max nr of predictors (at most 20)   15
Initial covariance computations
forward validation starts
forecast time point and code for predictors
24
25 42 41 40 39 38 26
26 42 41 40 39 38 26 7 6 22 14 23 28 15 11 2
27 42 41 40 39 38 26
28 42 41 40 39 38 26 25 29
29 42 41 40 39 38 26 25 29
30 42 41 40 39 38 26 25 29
31 42 41 40 39 38 26 25 29
32 42 41 40 39 38 26 25 29
33 42 41 40 39 38 26 25 29
34 42 41 40 39 38 26 25 29
35 42 41 40 39 38 26 25 34
36 42 41 40 39 38 26 25 36
37 42 41 40 39 38 34 22 6 7 28
38 42 41 40 39 38 34 22 6 7
39 42 41 40 39 38 34 22 6 7
40 42 41 40 39 38 34 29 6 7 20
41 42 41 40 39 38 34 29 6 7 20
42 42 41 40 39 38 34 29 6 7 16
43 42 41 40 39 38 10 34 29 6 7
44 42 41 40 39 38 34 29 6 7 9
45 42 41 40 39 38 34 29 6 7 9
46 42 41 40 39 38 34 29 6 7 9
47 42 41 40 39 38 34 29 6 7 9
****

Results
Selected model size incl season and trend 10 par.
Elimination of linear trend gives basic variance 879.699
trend + seasons give the variance 814.250
covariance estimated variance for selected model 170.643
** validation variance CMF for selected model *** 684.911
Do you want the variance functions? 1=yes, 0=no1

no param.	least squares	Akaike	decisionmeasure C(.)
5	814.250	1033.145	1116.774
6	379.165	504.559	794.883
7	314.600	439.059	779.959
8	257.860	377.424	719.993
9	190.725	292.775	707.220
10	170.643	274.724	610.879
11	155.329	262.265	720.762
12	138.484	245.226	888.789
13	124.139	230.546	1266.790
14	118.924	231.633	1636.879
15	112.185	229.163	1936.202

Do you want information about final model? 1=yes,0=no 1

coefficient	series	time lag
2.479	20	0
-0.385	10	4
2.202	3	0
-1.806	3	1
0.626	3	4

add
-1.66072 times predicted time
-268.871 at season 1
-326.821 at season 2
-287.209 at season 3
-321.275 at season 4
time=row number of data
season=season of predicted value
Residuals out? 1=yes, 0=no 1
time model sizes

i0-2,	i0-1,	i0,	i0+1,	i0+2,	chosen
24	50.53	49.09	54.27	55.99	46.60	0.00
25	30.18	33.06	29.59	19.17	8.07	55.56
26	30.79	37.25	40.10	48.05	68.57	59.44
27	13.11	17.75	7.17	20.94	23.78	27.08
28	7.04	-4.73	1.04	37.24	46.50	7.04
29	5.38	-13.29	-12.00	-16.99	-25.97	5.38
30	-17.11	-23.97	16.36	25.37	25.74	-17.11
31	-25.10	-27.37	-34.72	-23.73	-46.26	-25.10
32	-0.20	4.58	-9.91	-7.33	-8.58	-0.20
33	6.34	9.59	13.88	17.66	15.64	6.34
34	-35.11	-24.59	-20.43	-23.20	-26.97	-35.11
35	41.20	39.41	32.30	34.71	43.23	41.20
36	-41.08	-31.86	-30.18	-27.01	-21.39	-41.08
37	9.79	-4.32	-19.96	-15.59	-25.82	-19.96
38	38.41	7.52	6.98	9.36	9.54	7.52
39	40.42	39.45	28.96	27.98	23.59	39.45
40	-3.03	-17.67	-20.30	-20.55	-25.16	-20.30
41	-4.42	-18.39	-16.25	-22.56	-23.22	-16.25
42	-2.61	-0.37	2.16	2.95	-0.49	2.16
43	-57.86	-61.57	-45.52	-40.44	-30.69	-45.52
44	1.51	4.03	3.63	13.57	9.07	3.63
45	20.51	25.87	24.01	29.27	25.87	24.01
46	12.06	-15.11	-19.85	-29.38	-36.06	-19.85
47	4.49	12.74	15.81	15.95	12.63	15.81

4.4   Confusing information, large and small model sets

A human forecaster needs relevant information given as numerical data, drawn curves or the equivalent in order to make a sensible forecast. If the information becomes very extensive, he or she will have serious trouble in distinguishing essential information from inessential, and occasional covariations of a more or less random nature can be misjudged and give a poor forecast.

One may ask if a statistical analysis possibly runs the same risk. People in general seem to believe that information can never be harmful in such situations. However, the statistical methods we normally use are acting more as the human forecaster. If the information increases, for example because an increasing number of registered time series are becoming potential predictors in a multivariate model like (4.5), the result may well be a poor forecast compared to the case where only a few series were used. This overfitting problem has long been well known among statisticians, but methods to measure the effects have been lacking, especially so for the large model selection effects. Consequently there are strong reasons for trying to motivate which of the series really have a logical connection to the one predicted. Unfortunately this is not always easy or even possible. In many cases the series themselves are the only useful information we have. We may therefore also have good reasons to compare validation of a limited model set, based on a few selected time series, with validation when all the series are available in the model set.

The problem can be illustrated on a data base in the next section concerning paper and wood production and measured financial activities in various countries.

Are all series important for the prediction of series 1 in the data base, or how do we make a selection? The reader is encouraged to try fitting models of the type (4.5) using forward validation on different subsets of the data base. A program FORVAL.FOR can be used for this purpose. Anyone with a proper program for multivariate time series analyses can also program a forward validation routine around this according to the flow chart given earlier.

4.4.1 A multivariate data base

This data file contains 28 time series with quarterly data from 1972:1 to 1983:3. It thus has a 28-dimensional vector at 47 time points if no reduction is made.

Paper and board and related series, code for variables

1	SWE	TOTAL PAPER AND BOARD PRODUCTION
2	FIN	PRODUCTION, FUTURE TENDENCY % BAL
3	SWE	PRODUCTION, FUTURE TENDENCY % BAL
4	UK	CBI INTENSIONS INQUIRY, PLANT AND MACHINERY
5	SWE	PAPER, NEW EXPORT ORDERS
6	SWE	PAPER, DOMESTIC ORDERS
7	SWE	PAPER, VOLUME OF PRODUCTION
8	SWE	SAWMILLS, NEW EXPORT ORDERS
9	SWE	SAWMILL INDUSTRY, NEW DOMESTIC ORDERS
10	SWE	SAWMILL INDUSTRY, VOLUME OF PRODUCTION
11	SWE	PULP, INVENTORIES
12	SWE	PULP, CAPACITY UTILASITION %
13	SWE	PULP, NEW DOMESTIC ORDERS
14	SWE	PULP, NEW EXPORT ORDERS
15	SWE	PULP, PRODUCTION VOLUME
16	BEL	DOM. ORDERS INFLOW, TENDENCY % BAL.
17	FRA	PRODUCTION, FUTURE TENDENCY % BAL.
18	WG	GESCHÄFTSKLIMA VERARB. GEW.
19	US	NEW PRIVATE HOUSING UNITS STARTED
20	US	COMP. INDEX OF 12 LEADING INDICATORS, 1967=100
21	EG	PRODUCTION EXPECTATIONS %
22	WG	GESCHÄFTSKLIMA ZELLSTOFF, PAPIER UND P-ZEUG
23	CA	WOODPULP PRODUCTION, INDEX NUMBERS
24	CA	WOOD SUPPLY INDEX NUMBERS
25	FIN	NEWSPRINT PRODUCTION INDEX NUMBERS
26	FIN	WOODPULP PRODUCTION INDEX NUMBERS
27	FIN	WOOD SUPPLY INDEX NUMBERS
28	FIN	PAPER PRODUCTION INDEX NUMBERS

Series number

1	2	3	4	5	6	7	8	9	10	11	12	13	14	15	16	17	18	19	20	21	22	23	24	25	26	27	28
1972																											
1123	11	7	3	-7	-12	-22	19	-2	6	93	43	3	-39	-12	-11	21	-5	2334	122	5	9	74	71	93	96	44	85
1044	-13	10	9	38	16	-5	24	16	15	87	24	7	24	-25	1	23	0	2254	123	11	18	82	82	96	88	71	86
1138	36	14	16	33	14	12	36	3	5	75	33	6	18	5	-9	30	0	2481	128	16	14	68	71	99	85	99	104
1258	31	9	17	35	33	23	57	7	4	66	44	35	26	50	6	25	12	2366	131	20	27	79	75	120	94	91	102
1973																											
1319	55	17	30	66	38	44	18	2	15	-19	75	12	80	20	2	34	21	2365	134	22	30	82	77	112	78	63	94
1191	32	23	35	35	33	20	35	19	21	14	100	6	41	8	1	27	11	2067	133	24	19	88	92	52	37	35	58
1321	48	26	39	45	32	40	9	-6	5	-34	100	34	58	30	-7	30	-7	1874	131	19	23	61	69	95	80	99	106
1373	15	12	38	41	39	17	-19	-18	21	-22	100	19	20	11	-8	8	-28	1526	129	4	19	71	70	84	70	70	98
1974																											
1408	14	26	-3	21	12	12	-47	-19	16	-25	100	10	20	0	-1	14	-9	1555	128	2	35	86	73	90	69	54	104
1275	30	22	2	18	2	19	-74	-13	5	-46	100	6	20	2	-13	10	-16	1513	124	-2	37	84	65	66	55	59	89
1417	33	20	-6	-1	-8	9	-97	-33	-37	-37	100	0	0	-8	-24	-4	-34	1158	117	-22	-1	88	59	74	58	58	111
1410	23	1	-33	-44	-63	-19	-78	-49	-63	-46	100	-1	4	0	-40	-27	-37	975	109	-24	-37	89	42	86	64	40	90
1975																											
1252	-15	2	-41	-79	-72	-60	-23	-26	-71	10	80	-73	-56	-12	-47	-16	-32	993	108	-30	-58	77	45	60	56	27	74
1055	-14	-4	-27	-71	-74	-77	-30	-41	-43	88	27	-80	-99	-82	-25	-14	-36	1087	116	-25	-38	100	77	36	33	36	65
1068	-13	-6	-24	-33	-26	-3	-24	-37	-55	98	23	-31	-74	-11	-27	-8	-25	1264	121	-15	-30	48	38	49	35	48	81
1105	6	-16	-15	-11	-20	-35	47	4	2	99	19	-11	-41	-52	-9	18	-10	1321	123	9	-3	65	50	79	42	66	87
1976																											
1255	28	3	10	37	14	23	67	34	14	100	9	20	41	-14	4	28	3	1421	129	13	16	79	68	69	61	53	107
1172	27	9	15	40	34	23	11	16	11	97	38	14	44	-13	-4	25	2	1495	132	15	20	97	84	93	71	75	97
1244	-2	8	25	24	1	27	-8	-1	-14	95	27	0	-8	-9	-14	13	-8	1720	132	17	0	85	81	63	46	56	87
1306	-14	-17	26	-34	0	-22	-57	-32	-6	99	9	-5	-45	-71	-23	-10	-10	1804	135	11	-4	78	72	56	56	62	96
1977																											
1331	-8	-2	29	-36	-1	-25	9	4	-4	94	9	-1	3	-49	-19	5	-10	2063	138	2	-9	89	104	67	70	47	96
1188	7	-11	31	42	24	-6	-19	-15	5	96	4	-3	-13	-22	-19	-5	-17	1893	140	-5	-15	93	118	64	64	72	89
1243	17	-9	23	-5	-12	-19	4	-24	-23	100	4	-1	-60	-28	-35	-2	-20	1949	141	-3	-24	65	100	70	59	62	92
1299	-17	-11	21	-2	-23	1	-9	-39	-18	-57	29	-2	-24	4	-19	-11	-19	2142	142	-1	-6	90	98	72	76	64	97
1978																											
1363	-11	5	21	31	36	30	18	-8	16	-19	10	14	29	26	-11	11	-17	2032	145	0	8	86	93	60	82	62	106
1453	6	8	18	80	11	35	17	16	12	39	40	5	76	21	-13	11	-13	2070	147	0	9	99	116	90	83	84	90
1343	4	16	13	57	35	52	35	27	3	0	58	42	44	34	-17	6	-7	1970	147	5	17	92	107	80	83	88	102
1561	17	17	10	19	2	30	59	35	14	-39	85	9	35	45	-13	7	-1	2044	147	15	17	89	87	91	93	87	92
1979																											
1615	4	20	12	68	25	43	42	14	8	-68	88	23	31	-3	1	15	7	1847	149	10	30	117	108	124	93	59	101
1465	21	21	12	20	6	13	13	25	19	-79	90	4	20	7	4	10	5	1913	147	9	27	99	110	76	89	120	85
1549	29	21	8	-4	-2	15	10	10	3	-59	92	-2	-4	-3	2	10	0	1832	145	9	8	90	97	78	79	93	83
1652	22	6	-9	3	5	17	23	35	4	-60	93	9	-26	2	-2	8	-6	1498	141	7	10	93	74	94	105	100	101
1980																											
1683	14	11	-18	1	7	6	14	19	7	-70	89	11	26	-23	-7	11	0	1047	137	-1	-1	115	99	92	127	83	96
1309	7	6	-18	-26	-22	-29	-42	-29	-31	-25	98	1	4	-68	-17	-6	-20	1196	132	-14	-6	110	120	97	123	125	98
1585	-1	-3	-43	-75	-23	2	-84	-60	-17	-28	91	-9	-25	39	-28	-22	-27	1471	141	-24	-42	105	102	110	89	102	106
1605	-2	-17	-46	-87	-43	-57	-71	-57	-10	18	64	-66	-78	-24	-28	-20	-32	1482	143	-23	-22	93	82	105	118	93	106
1981																											
1632	4	-1	-38	-18	-24	-20	-70	-54	-41	41	59	4	-15	-11	-25	-10	-26	1305	142	-20	-7	105	111	107	103	68	93
1441	-9	-6	-30	19	2	-3	-67	-63	-33	16	53	0	1	-23	-14	-16	-28	1047	143	-15	-17	120	116	99	95	96	86
1489	10	-1	-23	-43	-29	28	-45	-45	-68	51	83	-8	-14	-17	-13	-10	-25	900	139	-7	-22	70	74	117	103	78	96
1561	-21	-6	-13	-53	-48	-33	36	-15	-32	40	53	-42	-40	-60	-13	-3	-25	906	136	-6	-22	90	75	82	86	88	96
1982																											
1608	-3	8	-5	-46	-27	-23	60	3	19	43	25	-53	-62	-56	-16	6	-24	920	135	-6	-17	102	94	99	105	65	106
1432	-12	-1	-2	-29	1	-10	24	-2	20	94	14	-3	-9	-10	-9	-7	-33	910	136	-7	-26	88	106	86	94	81	99
1375	-2	2	-11	-80	-13	-61	-14	-31	3	80	25	-21	-13	-49	-17	-19	-43	1134	138	-17	-29	79	98	94	58	63	100
1513	-9	-11	-20	-13	-10	-20	49	-15	11	66	41	-16	16	31	-4	-12	-32	1280	141	-17	-16	84	92	102	95	81	97
1983																											
1602	19	15	-5	22	9	6	55	4	44	21	52	8	14	52	-4	4	-10	1605	151	-4	13	106	107	96	108	69	113
1573	-8	11	6	23	0	8	28	15	27	30	74	3	39	6	-9	-19	-11	1736	157	-3	-6	101	124	103	74	76	100
1539	9	13	18	-1	3	25	6	1	1	40	77	-6	-8	24	-12	-23	-6	1664	160	2	11	91	110	103	91	66	117

4.5 Appendix

4.5.1 Examples of loss functions

All forecasts are not real numbers. They can be stated as vectors, probabilities or something else. Here we give some loss functions, which are applicable in different situations. Many more versions exist, but we do not strive for completeness. With computer intensive methods it is an easy task to evaluate all sorts of losses as long as they are possible to compute.

Scalar forecast $\hat{y}_t$ of scalar series y_t

$$L_1(\hat{y}_t, y_t) = |\hat{y}_t - y_t| \qquad \text{absolute error} \qquad (4.11)$$

$$L_2(\hat{y}_t, y_t) = (\hat{y}_t - y_t)^2 \qquad \text{squared error.} \qquad (4.12)$$

$$L_3(\hat{y}_t, y_t) = \begin{cases} (\hat{y}_t - y_t)^2 & \text{if } |\hat{y}_t - y_t| \le a \\ 2a|\hat{y}_t - y_t| - a^2 & \text{if } |\hat{y}_t - y_t| > a. \end{cases} \qquad (4.13)$$

The squared loss is by far the most widely used and leads in standard inference situations to estimates of the average type. The absolute error is robust against large errors and leads to inferences of the median type in some standard situations. In situations where we suspect miscoded data and other difficulties leading to very large errors, robustness is valuable and can replace some careful and otherwise necessary data inspection. However, absolute error estimation is analytically more complex than least squares methods. The measure L_3 is Huber's compromise (Huber, 1964). It keeps some of the advantages of the least squares methods for ordinary errors (if a is set as the limit of the ordinary) but gives robustness against large errors.

Vector valued forecast $\hat{\mathbf{y}}_t$ of vector valued $\mathbf{y}_t$

$$L_4(\hat{\mathbf{y}}_t, \mathbf{y}_t) = (\hat{\mathbf{y}}_t - \mathbf{y}_t)'\mathbf{D}(\hat{\mathbf{y}}_t - \mathbf{y}_t), \qquad (4.14)$$

where $\mathbf{D}$ is a positive definite matrix. The special case

$$D = \begin{pmatrix} d_1 & & 0 \\ & \ddots & \\ 0 & & d_m \end{pmatrix}$$

gives $L_4 = \sum d_i(\hat{y}_{it} - y_{it})^2$ where y_{it} is the ith component of the vector $\mathbf{y}_t$.

Predictive distribution

The single valued forecast $\hat{y}_t$ can be replaced by a predictive probability distribution with density function $\hat{f}_t(y)$. This can for example be the model's conditional distribution of y_t given the preceding observations. Put

$$L_5(\hat{y}_t, y_t) = L_5(\hat{f}_t(\cdot), y_t) = -\ln \hat{f}_t(y_t). \qquad (4.15)$$

Here y_t may be either scalar or vector valued.

This measure means that a large logarithm of the density is wanted where the observation comes. Compare this measure with the log likelihood used for maximum likelihood estimation of parameters.

A particular case of a predictive distribution is when the forecast gives the probability of an event. If $y_t = 1$ or 0 according to whether it rains day t or not, a forecast $\hat{y}_t$ could be a number $0 < \hat{y} < 1$ giving the model's probability of $y_t = 1$. The measure (4.15) then becomes

$$L_6(\hat{y}_t, y_t) = \begin{cases} \ln \hat{y}_t & \text{if } y_t = 1 \\ \ln(1 - \hat{y}_t) & \text{if } y_t = 0. \end{cases} \qquad (4.16)$$

This can be summarized as $L_6 = y_t \ln \hat{y}_t + (1 - y_t)\ln(1 - \hat{y}_t)$.

An alternative measure in this situation could be

$$L(\hat{y}_t, y_t) = (\hat{y}_t - y_t)^2$$

which is simply (4.12). The measure (4.16) is more sensitive for values $\hat{y}_t$ close to 0 and 1 than (4.12). If precision of low probabilities is important, the logarithmic measure is therefore preferred.

4.6 Forward validation FORTRAN code

```fortran
c       Forward validation of time series model selection
c       designed for examples on uni- or multivariate
c       time series with trend and seasonal effects.
c       Mathematical Statistics, Linkoping, Urban Hjorth.
c       administrative arrays for the validation
        dimension d(20),v(20),c(20),res(249,21)
c       regression arrays
        dimension xx(200,200),ipred(200),iforce(20),jser(20)
     1      ,predik(20,20),prefel(20)
c       arrays for transformation of the data base
        dimension x(249,34),z(249,200),isx(34),ipt(20),xm(34),b(34)
        character*30 datfil
        write(*,*)' '
        write(*,*)' Forward validation of time series. '
        write(*,*)' Limitations:   max 249 time points'
        write(*,*)'                max 34 series'
        write(*,*)'                quadratic or linear trend'
        write(*,*)'                max 199 predictor candidates'
        write(*,*)'                max 20 selected prdictors'
        write(*,*)'        (can be adjusted to memory limits)'
        write(*,*)' '
        write(*,*)' Row one of the data file must have'
        write(*,*)' the following five numbers'
        write(*,*)' nr of series (columns), nr of time points'
        write(*,*)' (rows below the first), nr of seasons, '
        write(*,*)' year (block), and season of first data'
        write(*,*)' '
        write(*,*)' Give the name of your data file,'
        read(*,2)datfil
2       format(A)
        open(22,file=datfil)
        read(22,*)nser,ntid,nsea,ny1,nsea1
        nsea=max(1,nsea)
        write(*,*)' '
        write(*,*)'The file has'
        write(*,*) nser,' series observed at'
        write(*,*) ntid,' time points. Times are divided into'
        write(*,*) nsea,' seasons.'
        write(*,*)' First year (block) and season  ',ny1,nsea1
        do 10 j=1,ntid
10      read(22,*)(x(j,i),i=1,nser)
c       transform to numerically more stable data
        an=float(ntid)
        do 20 i=1,nser
        xm(i)=0.
        xt=0.
        tt=0.
        do 30 j=1,ntid
        t=float(j)-(an+1.)/2.
        xm(i)=xm(i)+x(j,i)
        xt=xt+x(j,i)*t
30      tt=tt+t*t
        xm(i)=xm(i)/an
        b(i)=xt/tt
        do 40 j=1,ntid
        t=float(j)-(an+1.)/2.
40      x(j,i)=x(j,i)-xm(i)-b(i)*t
20      continue
45      write(*,*)' '
        write(*,*)' Which series (column) is predicted?'
        read(*,*) isy
        write(*,*)' How many time-steps ahead?'
        read(*,*) ktau
        write(*,*)' How many series are used as predictors?'
        read(*,*) nsx
```

```
        write(*,*)' Give the numbers of the predictor series'
        read(*,*) (isx(k),k=1,nsx)
        write(*,*)' How many different predictor time lags?'
        read(*,*) npt
        write(*,*)' Give predictor lags (ages), 0=present time'
        read(*,*) (ipt(k),k=1,npt)
        mlag=0
        do 50 k=1,npt
50      mlag=max0(mlag,ipt(k))
        nobs=ntid-ktau-mlag
        do 200 j=1,nobs
        z(j,1)=x(ktau+j+mlag,isy)
        index=1
        do 210 k=1,nsx
        do 210 l=1,npt
        index=index+1
210     z(j,index)=x(j+mlag-ipt(l),isx(k))
200     continue
c       season and trend variables
        do 230 ks=1,nsea
        index=index+1
        do 230 j=ks,nobs,nsea
230     z(j,index)=1.
        index=index+1
        do 240 j=1,nobs
        t=float(2*j-1-nobs)/2.
240     z(j,index)=t
        write(*,*)' Quadratic trend? 1=yes, 0=no'
        read(*,*) kv
        if(kv.le.0)goto 260
        index=index+1
        do 250 j=1,nobs
        t=float(2*j-1-nobs)/2.
250     z(j,index)=t*t
260     ndim=index
        if(ndim.gt.200)write(*,*)'Over 199 predictors, new try'
        if(ndim.gt.200)goto 45
        write(*,*)' '
        write(*,*)' Season and trend are forced into models'
        nforce=nsea+2
        if(kv.le.0)nforce=nsea+1
        do 330 i=1,nforce
330     iforce(i)=ndim+1-i
        write(*,*) ndim-1,' predictors. ',nforce,' are forced'
        write(*,*)' Minimum ',nforce,' predictors in models,'
        write(*,*)' Give max nr of predictors (at most 20)'
        read(*,*) ipmax
        write(*,*)' '
        write(*,*)' Initial covariance computations'
        im=ipmax+4
        do 400 i=1,ndim
        do 400 j=1,ndim
400     xx(i,j)=0.
        do 410 it=1,im-1
        do 410 i=1,ndim
        do 410 j=1,ndim
410     xx(i,j)=xx(i,j)+z(it,i)*z(it,j)
        iy=1
        np=ndim-1
        iter=ipmax
        do 405 i=1,np
405     ipred(i)=i+1
        write(*,*)' Forward validation starts'
        write(*,*)' '
        write(*,*)' Forecast time and code for predictors'
        do 500 it=im,nobs
        t=float(it)
        call fsreg(xx,iy,ipred,np,iforce,nforce,iter,jser
     1       ,predik,prefel)
```

```
         write(*,96) it+ktau+mlag, (jser(i), i=1,i0)
 96      format(i3,2x,20i3)
         do 510 ip=nforce,ipmax
         p=float(ip)
 c       prognosis nr ip at time it
         ytak=0.
         do 520 i=1,ip
 520     ytak=ytak+predik(ip,i)*z(it,jser(i))
         res(it,ip)=z(it,1)-ytak
         if(ip.eq.i0)res(it,21)=res(it,ip)
 c       loss function
         al=(z(it,1)-ytak)**2
 c       weights
         g=1./(1.+p/(t-1.))
         d(ip)=d(ip)+g
         v(ip)=v(ip)+g*al
         c(ip)=v(ip)/d(ip)
         if(ip.eq.i0.and.it.gt.im)
 1          cn=cn+g*al*(1.+p/float(nobs))/float(nobs-im-1+3)
 510     continue
 c       selection of i0
         amin=.9e32
         do 530 ip=nforce,ipmax
         if(amin.lt.c(ip)) goto 530
         amin=c(ip)
         i0=ip
 530     continue
 c       updating the xx-matrix
         do 540 i=1,ndim
         do 540 j=1,ndim
 540     xx(i,j)=xx(i,j)+z(it,i)*z(it,j)
 500     continue
         write(*,*)'**********************************************'
         write(*,*)' '
         write(*,*)'                     Results'
         write(*,*)' '
         write(*,*)' Selected model size         ',i0,' par.'
         call fsreg(xx,iy,ipred,np,iforce,nforce,i0,jser,predik,prefel)
         write(*,*)' '
         gv=xx(1,1)/float(nobs)
         tsv=prefel(nforce)/float(nobs)
         write(*,*)' Fitting linear trend gives the basic variance ',gv
         write(*,*)' trend + seasons give the variance        ',tsv
         write(*,*)' covariance est variance for selected model'
 1          ,prefel(i0)/float(nobs)
         write(*,*)' ** Validation variance CMF for sel model ** ',cn
         write(*,*)' '
         write(*,*)' Do you want variance functions? 1=yes, 0=no'
         read(*,*)mm
         if(mm.ne.1)goto 600
         write(*,*)' No parameters, least squares,    Akaike,
 1          decision measure C(.)'
         do 560 ip=nforce,ipmax
         fn=float(nobs)
         p=float(ip)
         sq=prefel(ip)/fn
         aic=sq*exp(2.*p/fn)
 560     write(*,99) ip,sq,aic,c(ip)
 99      format(i9,f18.3,f14.3,f18.3)
 600     continue
         write(*,*)' '
         write(*,*)' Do you want info about final model? 1=yes,0=no'
         read(*,*) inf
         if(inf.ne.1) goto 650
         write(*,*)' '
         write(*,*)' coefficient  series  time lag'
         con=0.
         trl=0.
```

```
         tm=(an+1.)/2.
         if(i0.eq.nforce) goto 630
         do 610 i=nforce+1,i0
         nr=1+int(float(jser(i)-2)/float(npt)+.00001)
         il=jser(i)-1-npt*(nr-1)
         write(*,98) predik(i0,i),isx(nr),ipt(il)
         con=con+predik(i0,i)*(-xm(isx(nr))
   1        +b(isx(nr))*(ktau+ipt(il)+tm))
         tr1=tr1-predik(i0,i)*b(isx(nr))
 98      format(f10.3,i6,i6)
 610     continue
         write(*,*)' add'
 630     continue
         con=con+xm(isy)-b(isy)*tm
         tr1=tr1+b(isy)
         ttm=float(2*ntid-nobs-1)/2.
         if(kv.le.0)then
         tr2=0.
         slope=predik(i0,1)
         elseif(kv.gt.0)then
         tr2=predik(i0,1)
         slope=predik(i0,2)
         endif
         con=con-slope*(ktau+ttm)+tr2*(ktau+ttm)**2
         tr1=tr1+slope-2*tr2*(ktau+ttm)
         write(*,*) tr1,'    times predicted time'
         if(kv.gt.0) write(*,*) tr2,'   times squared pred time'
         if(nsea.eq.1) then
         write(*,*) predik(i0,nforce)+con,' constant'
         elseif(nsea.gt.1) then
         do 642 i=1,nsea
 642     if(mod(nsea1+ntid-nobs+i-1,nsea).eq.1) is1=i
         ks=1
         do 645 j=is1,nsea
         write(*,*) predik(i0,nforce+1-j)+con,'   at season',ks
 645     ks=ks+1
         if(is1.gt.1) then
         do 646 j=1,is1-1
         write(*,*) predik(i0,nforce+1-j)+con,'   at season',ks
 646     ks=ks+1
         endif
         write(*,*)'              time=row number of data'
         write(*,*)'              season=season of predicted value'
         endif
         write(*,*)'   '
 650     write(*,*)' Residuals out? 1=yes, 0=no'
         read (*,*) itt
         if(itt.lt.1) goto 700
         write(*,*)' time,  model size',' i0-2,i0-1,i0,i0+1,i0+2,','chosen'
         do 640 it=im,nobs
         write(*,97) it+ktau+mlag,(res(it,j),j=i0-2,i0+2),res(it,21)
 640     continue
 97      format(i4,5f10.2,f15.2)
 700     continue
         end
         subroutine fsreg(r,iy,ipred,np,iforce,nforce,iter,jser,predik
   1     ,prefel)
         dimension r(200,200),ipred(200),jser(20),predik(20,20)
         dimension iforce(20),f(20),p(20,20)
         dimension prefel(20), a(20,20),b(20,20),c(20),d(20),e(20,20)
  c      singularity guard not included
         do 100 i=1,iter
         do 100 j=1,iter
         a(i,j)=0.
         b(i,j)=0.
         if(i.eq.j) b(i,j)=1.
         predik(i,j)=0.
 100        continue
```

```
c     step 1
      test=0.
      do 120 i1=1,np
      i=ipred(i1)
      if(nforce.ge.1) i=iforce(1)
      h=r(i,i)
      g=r(iy,i)
      x=g*g/h-test
      if(x.le.0.) goto 120
      c(1)=g
      d(1)=h
      jser(1)=i
      test=g*g/h
      if(nforce.ge.1) goto 122
120      continue
122      continue
c     step 2 to iter
      do 150 n=2,iter
      n1=n-1
      test=0.
      do 160 i1=1,np
      i=ipred(i1)
      if(nforce.ge.n) i=iforce(n)
      do 180 k=1,n1
180      if (i.eq.jser(k)) goto 160
      do 200 k=1,n1
200      f(n-k)=r(i,jser(k))
      do 190 k=1,n1
190      a(n,k)=0.
      do 230 k=1,n1
      do 230 j=1,k
230      a(n,k)=a(n,k)+b(k,j)*f(n-j)
      do 250 k=1,n1
250      a(n,k)=-a(n,k)/d(k)
      g=0.
      h=0.
      do 260 k=1,n1
      g=g+a(n,k)*c(k)
260      h=h+a(n,k)**2*d(k)
      g=g+r(iy,i)
      h=r(i,i)-h
      x=g**2/h
      if(x.le.test) goto 160
      test=x
      c(n)=g
      d(n)=h
      jser(n)=i
      do 280 k=1,n1
280      e(n,k)=a(n,k)
      if(nforce.ge.n) goto 162
160      continue
162      continue
      do 300 k=1,n1
      do 300 j=k,n1
300      b(n,k)=b(n,k)+e(n,j)*b(j,k)
150      continue
      do 320 i=1,iter
      do 320 j=1,i
320      p(i,j)=c(j)/d(j)
      do 350 i=1,iter
      do 350 j=1,iter
      do 350 k=1,iter
350      predik(i,j)=predik(i,j)+p(i,k)*b(k,j)
      do 360 i=1,iter
      prefel(i)=r(iy,iy)
      do 360 j=1,i
360      prefel(i)=prefel(i)-c(j)**2/d(j)
      return
      end
```

4.7 Exercises

The following exercises presume the reader has a working version of a forward validation program on the computer.

Exercise 4.1 Do forward validation on the prediction of series one in the paper and board data set. Use as predictors lags 0 to 6 of the same series one. Predict both one time step ahead and three time steps ahead. Is there any difference in the selection based on Akaike's criterion and the validation criterion?

Exercise 4.2 Write down the class of models used in Exercise 4.1 as simply and explicitly as possible.

Exercise 4.3 Repeat the forward validation of series one, but use a selected set of series as predictors. Try to find a good set by considering the importance of competition from Finland, Canada (when the US market does not absorb their production), and the main European customers in WG, UK, and FRA etc. For the comparison of different runs, use the same maximal lag and the same maximal model size so that the validation is performed on exactly the same subsequence of the data. Compare the different measures of mean square prediction errors.

Exercise 4.4 See if all series offered as predictors give a better or worse validation result for the same predictions as in the previous examples.

Exercise 4.5 Suppose that after some experimentation you have found a set of predictors that give a very low validation error. Decide if there is any danger in using this result. Have you validated the model selection procedure which is actually used?

CHAPTER 5

Statistical bootstrap

The word bootstrap hints at the saying 'pull oneself up by ones bootstraps', which is rather close to doing the impossible. This was perhaps what Bradley Efron felt he had brought about when introducing his new idea in the *Annals of Mathematical Statistics*, 1979.

The bootstrap technique is a very general method to create measures of uncertainty and bias, in particular at parameter estimation from independent identically distributed variables. However, the method has been generalized much further to attack problems like discrimination, where objects are sorted into different classes, regression problems where the data are independent but not identically distributed, and certain problems for stochastic processes where the data are highly dependent.

On the surface, the bootstrap idea may seem simple. In fact, like most good ideas it is not very complicated, and the method might well have come up in the early days of statistics if it had not been so dependent on computer simulations. But the selection of this particular method among all possible ways to use the computer, and the recognition of its capacity is built on a very deep feeling for statistical inference.

5.1 The parameter concept

The original bootstrap is distribution free, which means that it is not dependent on a particular class of distributions. It just requires a proposed estimator $\hat{\theta}$ of a parameter θ. To find this estimator $\hat{\theta}$ is a problem left entirely outside the bootstrap theory, but sometimes a bootstrap analysis can motivate an adjusted estimator. There is just one problem left. The parameter θ must have a definition which can be interpreted for a wide class of distribution functions. This is so because in the analysis, the parameter will be defined not only in our model of reality but also when this reality in a certain sense is replaced by the observations taken on it. Two versions of an example can illustrate this.

Example 5.1a   Median estimation, parameter definition
Let $x_1, \ldots, x_{11}$ be independent observations from a continuous distribution $F(x)$. Define the parameter θ as the median of $F(x)$. We suppose that $F(\theta) = 1/2$ has a unique solution θ.

Let $x_{(1)} < x_{(2)} < \ldots < x_{(11)}$ be the ordered sample. The most natural estimate of the parameter θ will be the median of our data so we use

$$\hat{\theta}_a = x_{(6)}. \tag{5.1}$$

Now let reality, represented by $F(x)$, be replaced by the observations $x_{(1)} < x_{(2)} < \ldots < x_{(11)}$. This is performed in the following way. Represent these data by the empirical distribution function $F_n(x)$, $n = 11$, putting probability $1/n$ on each observed value.

$$F_n(x) = \frac{\#\{x_i \leq x\}}{n} = \begin{cases} 0 & x < x_{(1)} \\ k/n & x_{(k)} \leq x < x_{(k+1)}, \ k < n \\ 1 & x_{(n)} \leq x. \end{cases}$$

The theoretical median is now replaced by the same parameter in $F_n(x)$. We denote this by a new symbol $\tilde{\theta}$ and get of course

$$\tilde{\theta} = x_{(6)}. \tag{5.2}$$

The fact that the new median (5.2) and the most natural estimate (5.1) are equal is an accidental occurrence, but it happens now and then. The basic condition on the parameters is that they are clearly defined by the same functional of the distribution functions F and F_n respectively

$$\theta = g(F(\cdot))$$
$$\tilde{\theta} = g(F_n(\cdot)). \qquad (5.3)$$

Sometimes we prefer the shorter notation $\theta = \theta(F)$, $\tilde{\theta} = \theta(F_n)$. In situations when the same estimate $\hat{\theta} = \theta(F_n)$ is used, as in (5.1), we get $\hat{\theta} = \tilde{\theta}$.

Example 5.1b   A different estimate of the median, parameter definition
If the distribution $F(x)$ is symmetric around its median θ, and the expected value $\mu = \int_{-\infty}^{\infty} x \, dF(x)$ exists, then μ and the median θ are identical. The expected value is usually estimated by the mean $\bar{x}$ of the observations. When the data are normally distributed it can be shown that this mean is an optimal estimator of μ, according to all reasonable symmetric criteria of optimality. Since $\theta = \mu$, the mean also becomes optimal for estimating θ in this and similar cases. It is thus quite sensible under some assumptions to estimate the median by the mean $\bar{x}$. When $F(x)$ is non-symmetric, the mean is usually biased and loses interest as an estimator of the median. Nevertheless, it can be interesting here to consider the parameter definition, and to study the effects of bootstrapping and bias correcting such an estimate as we will do later in Section 5.4.2. Let therefore

$$\hat{\theta}_b = \bar{x}. \qquad (5.4)$$

The definition of the parameter $\tilde{\theta}$ in the empirical distribution is not affected by this change of estimator. We still define θ as the median of $F(x)$ and $\tilde{\theta}$ as the median of $F_n(x)$, which means that $\tilde{\theta} = x_{(6)}$.

Limitation of the parameter definition
It is a limitation that the parameter should be defined both in $F(x)$ and in the empirical distribution $F_n(x)$ (condition (5.3)). The parameters α, β in the Weibull distribution with

density function

$$f(x) = \alpha\beta x^{\beta-1}e^{-\alpha x^{\beta}}, \quad x > 0$$

have for example no immediate interpretation in the empirical distribution. Nevertheless something can be done in this situation. We may define other parameters in the Weibull distribution such as the expected value and the variance, or two percentiles. These new parameters will have a one-one relation to α and β and they exist also in the empirical distribution. We may also perform a bootstrap analysis without any defined parameter in F_n, and get valuable measures on the variability of an estimator, but then we are not able to observe the bias or make any bias correction or a confidence interval for the parameter.

Asymptotic bootstrap analysis is often restricted to the estimator $\hat{\theta} = \theta(F_n)$, which rules out the estimator $\hat{\theta}_b$ for the median and makes $\tilde{\theta} = \hat{\theta}$ by definition. See Hall (1992) for such an exposition. Here we use a more general setting.

5.2 Classical measures and the bootstrap

Let $x_1, \ldots, x_n$ be n observations of an object, taken in such a way that in the model they are independent random variables with a common distribution $F(x)$. Let θ be a real valued, scalar parameter which we wish to estimate by a function of the observations $\hat{\theta} = \hat{\theta}(x_1, \ldots, x_n)$. The expected value of the estimator is given by the integral

$$E_F[\hat{\theta}] = \int \ldots \int \hat{\theta}(x_1, \ldots, x_n)f(x_1)\cdots f(x_n)\, dx_1 \ldots dx_n$$

if $F(x)$ has the derivative $f(x)$. Otherwise replace $f(x_i)dx_i$ with the Stieltjes symbol $dF(x_i)$ to cover sums and the general case. The notation E_F indicates that the expected value is computed with the distribution F. The Bias of this estimator is defined accordingly as

$$\text{Bias} = E_F[\hat{\theta} - \theta] \qquad (5.5)$$

and can be a function of the parameter θ (and of possible other parameters in F). The estimator $\hat{\theta}$ is unbiased if $E_F[\hat{\theta} - \theta] = 0$ for all θ, or in case of more parameters for all parameter vectors considered in the model. The variance of $\hat{\theta}$ is given by

$$\text{Var}(\hat{\theta}) = E_F[(\hat{\theta} - E_F[\hat{\theta}])^2] \qquad (5.6)$$

and the standard deviation $\sigma_{\hat{\theta}}$ is the square root of the variance as usual. More detailed information of $\hat{\theta}$ is achieved by studying the distribution of $\hat{\theta}$ or of $\hat{\theta} - \theta$, $\hat{\theta}/\theta$, or some other convenient expression for comparing the estimator and the parameter. This is required when we compute confidence intervals for θ. We will mainly discuss differences $\hat{\theta} - \theta$, like most of the bootstrap literature does, but sometimes the ratio $\hat{\theta}/\theta$ has a more stable distribution. If this is the case, and both the parameter and the estimate are always positive, we may as well study $\log(\hat{\theta}/\theta) = \log\hat{\theta} - \log\theta$ instead. If we transform to a new parameter $\nu = \log\theta$ and a corresponding estimate $\hat{\nu} = \log\hat{\theta}$, this situation is transformed into differences. More generally, the construction of pivot variables with approximately parameter independent distributions is very useful in bootstrap analyses, and we will return to that issue later on.

When parameters are estimated, the distribution F is unknown. At least it has an unknown parameter θ, for why else should we estimate it. The distribution of the estimator, and in particular the expected values (5.5) and (5.6), are defined by this unknown distribution. They are therefore as a rule impossible to compute exactly. Statistical inference methods try different ways to get around this basic problem, and *the idea of the bootstrap method is to replace the unknown distribution $F(x)$ by the empirical distribution $F_n(x)$ in (5.5), (5.6) and all other computations of interest.* This is spelled out in more detail in the following sections.

Let us consider some classical alternatives. If $F(x; \theta)$ is the class of univariate normal distributions, and a sample $X_1, \ldots, X_n$ is available, one can circumvent the problem for the basic parameters by studying quantities with distributions that are independent of the parameter. This is the best known example of pivot variables. It is well known that the variable

$$U = \frac{\bar{X} - \mu}{s/\sqrt{n}}$$

is t-distributed with $n-1$ degrees of freedom, no matter what values the parameters μ or σ have. Furthermore $(n-1)s^2/\sigma^2$ is χ^2-distributed with $n-1$ degrees of freedom regardless of the value of σ. This can be used for the construction of confidence intervals for μ and σ.

Many other parametric problems can be handled in a similar way. Often convergence to the normal distribution is motivated by the central limit theorem when sample sizes get large, and this is useful for deriving approximations. By this technique we can approximate the distributions of the parameter estimates for many well known distributions such as the Poisson, Binomial, Exponential, etc. However, many situations are less well defined. A good thing with the bootstrap method is that it is not limited to known parametric distributions and analytically tractable situations.

5.3   The bootstrap method

In order to explain the bootstrap method we will think of one real and one artificial problem which are very similar. The artificial one is called the bootstrap problem.

In *the real problem* we have data $x_1, \ldots, x_n$ and a model which only states that these data are independent observations from some distribution $F(x)$. Very weak assumptions about F or none at all are made. Perhaps we assume it is continuous. (We talk about the original bootstrap only, generalizations will come later.) A parameter $\theta = g(F(.))$ is defined (in principle) by the distribution and is estimated by $\hat{\theta} = \hat{\theta}(x_1, \ldots, x_n)$. This estimator is regarded as already given intuitively or from some other theory.

The *bootstrap problem* imitates the real problem but with F replaced by the empirical distribution $F_n(x)$ and the parameter by that of the empirical distribution,

$$F_n(x) = \frac{\#\{x_i \le x\}}{n}, \qquad \tilde{\theta} = g(F_n(.)). \qquad (5.7)$$

These quantities are known since the data $x_1, \ldots, x_n$ have been observed. We now simulate the same number of independent observations from $F_n(x)$ and denote this *bootstrap sample* as

$$X_1^*, X_2^*, \ldots, X_n^*, \quad \text{where } X_i^* \in F_n(x). \qquad (5.8)$$

Compute the parameter estimate

$$\hat{\theta}^* = \hat{\theta}(X_1^*, \ldots, X_n^*) \qquad (5.9)$$

by the same estimator as in the real problem.

Repeated simulations of (5.8) and (5.9) will show how accurately $\hat{\theta}^*$ estimates $\tilde{\theta}$ in the bootstrap problem. In particular we get simulated approximations of

$$E_*[\hat{\theta}^* - \tilde{\theta}] = \text{Bias}^*$$
$$\text{Var}_*(\hat{\theta}^*), \qquad (5.10)$$

and of the distribution of the errors

$$P_*(\hat{\theta}^* - \tilde{\theta} \le u). \qquad (5.11)$$

The index * denotes that expected values and probabilities only consider variations in the bootstrap simulations, while the data $x_1, \ldots, x_n$ are held fixed at their observed values. The correct interpretation of the starred quantities is therefore that they are conditional expected values and probabilities given the observations $x_1, \ldots, x_n$. Since we have a one-one relation between the set of observed values and the empirical distribution $F_n(x)$, we can also say that E_* and P_* are conditional on $F_n(x)$,

$$E_*[\hat{\theta}^* - \tilde{\theta}] = E[\hat{\theta}^* - \tilde{\theta}|F_n(x)].$$

The bold conclusion is now that *estimation properties of the bootstrap problem can be used to judge the properties of the real estimation problem.*

Using this idea, the variability of $\hat{\theta}$ can be estimated by $\sigma_* = \sqrt{\mathrm{Var}_*(\hat{\theta}^*)}$, or by a histogram or some other picture of the distribution of $\hat{\theta}^* - \tilde{\theta}$. If the bias is large, a bias corrected estimate can be given as $\hat{\theta}_1 = \hat{\theta} - \mathrm{Bias}^*$.

This is simply another function of the original data, provided the bootstrap simulations are large enough to find the expected value Bias* with accuracy. The Bias* is namely defined by the empirical distribution which is a function of the original data.

5.3.1 Flow chart of bootstrapped parameter estimation

We summarize the contents of the last section by a flow chart showing the important steps in a bootstrap analysis of parameter estimation from n independent identically distributed observations.

INITIATION

$x_1, \ldots, x_n$ iid $F(x)$

$\hat{\theta} = \hat{\theta}(x_1, \ldots, x_n)$

$\tilde{\theta} = g(F_n(x))$ from $\theta = g(F(x))$

BOOTSTRAP LOOP

$i = 1, \ldots, NB$

RESAMPLE

$j = 1, \ldots, n$

$k = \text{int}(1 + n \star \text{random}(\cdot))$

$X_j^\star = x_k$

end of j-loop

$\hat{\theta}_i^\star = \hat{\theta}(X_1^\star, \ldots, X_n^\star)$

end of i-loop

RESULTS, OUTPUT

$\text{AVE} = \Sigma \hat{\theta}_i^\star / NB$

$\text{BIAS}^\star = \text{AVE} - \tilde{\theta}$

$\text{STD} = \sqrt{\Sigma(\hat{\theta}_i^\star - \text{AVE})^2 / (NB - 1)}$

$\hat{\theta}_1 = \hat{\theta} - \text{BIAS}^\star$

HISTOGRAM OF $\hat{\theta}^\star$ OR $\hat{\theta}^\star - \tilde{\theta}$

CONFIDENCE INTERVAL FOR θ

5.4   Numerical illustration in two versions

5.4.1   Median estimation by the median

Eleven life lengths of an engine part were measured as

5700, 36300, 12400, 28000, 19300, 21500, 12900, 4100, 91400, 7600, 1600.

We regard the values as independent observations x_i on a continuous distribution $F(x)$ with median θ. Estimate θ by the median of the data

$$\hat{\theta} = x_{(6)} = 12900.$$

Bootstrap simulations

To draw a bootstrap sample from $F_n(x)$ is equivalent to drawing each X_i^* at random among the observed values x_1, $\ldots, x_{11}$. Since X_i^* are independent (given $F_n(x)$), we draw the observations with replacement, and the same value can be taken more than once. In Table 5.1 the values drawn in the first five simulations are indicated by +, and the corresponding parameter estimates computed. The true parameter value in $F_n(x)$ is $\tilde{\theta} = 12900$.

These initial simulations give an average of $\hat{\theta}^*$ which is 13460. This is 560 units above the parameter value $\tilde{\theta} = 12900$ in the distribution $F_n(x)$. The standard deviation of the five simulated values of $\hat{\theta}^*$ is 5018. Let us now look at the results of some more simulations. The diagram in Figure 5.1 shows 200 simulated values of $\hat{\theta}^*$.

The 200 simulations of $\hat{\theta}^*$ gave the estimates

average	$E_*[\hat{\theta}^*] = 14843$
variance	$\text{Var}_*(\hat{\theta}^*) = 32.9 \cdot 10^6$
standard deviation	$s_* = 5737$
bias estimate	$\text{Bias}^* = 14843 - 12900 = 1943.$

Extending the simulation to 2000 bootstrap samples gave the average 15085 and the standard deviation 5732 which means relatively minor adjustments compared to the results after 200 simulations.

Table 5.1 *Data drawn in the first five bootstrap samples.*

Original data ordered	Bootstrap sample number				
	1	2	3	4	5
1600			+		+ +
4100	+ + +	+ +	+	+	
5700	+	+	+	+ + +	+
7600				+ +	+ +
12400	+	+	+		+
12900	+		+ +		
19300	+	+	+	+ +	
21500		+ + +	+	+	+
28000	+	+ +	+	+ +	+
36300	+	+			+
91400	+ +		+ +		+ +
$\hat{\theta}^*$	12900	21500	12900	7600	12400

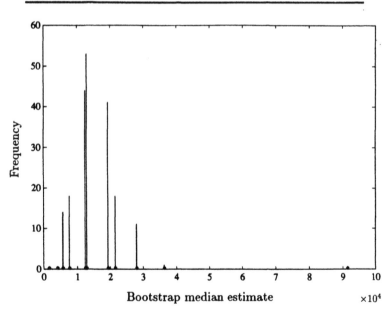

Figure 5.1 *Bar chart of 200 independent bootstrap simulations. Bullets on the x-axis show the original data.*

Conclusions from 200 simulations
We will now draw the following conclusions from the boot-strap problem over to the real problem. The original param-eter estimate $\hat{\theta} = 12900$ has
  a. an estimated bias of 1943 units,
  b. an estimated standard deviation $\hat{s}_{\hat{\theta}} = 5737$,
  c. a distribution somewhat like the bar chart above (but continuous).

A bias adjusted estimate is given by
  d. $\hat{\theta}_1 = 12900 - 1943 = 10957$.

Notice in particular that although the average bootstrap re-sult lies above 12900, the conclusion is that the estimate is adjusted down below 12900. If our bootstrap estimation is on average too high, then the real estimate is probably too high as well.

5.4.2 Median estimation by the average

We will next study what happens if, for some reason, we choose the average $\bar{x}$ as an estimator of the median and then apply the bootstrap technique. Motivations for this estima-tor were mentioned in Example 5.1b. We use the same eleven data points as in Section 5.4.1, namely 5700, 36300, 12400, 28000, 19300, 21500, 12900, 4100, 91400, 7600, 1600. Let θ be the median of the theoretical distribution $F(x)$. Our estimator now becomes

$$\hat{\theta}_b = \bar{x} = 21891.$$

This value is far away from the earlier estimates.
  The bootstrap problem starts as before with the empiri-cal distribution $F_n(x)$ and its median $\tilde{\theta} = 12900$. In each simulation, eleven independent observations are drawn from $F_n(x)$. The bootstrap estimate

$$\hat{\theta}_b^* = \frac{X_1^* + \ldots + X_{11}^*}{11}$$

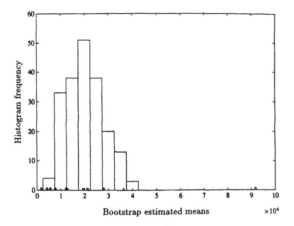

Figure 5.2 *Histogram of estimates $\hat{\theta}_b^*$ in 200 bootstrap simulations. Original data are shown as bullets.*

is then computed. The result of 200 such simulations are drawn as a histogram in Figure 5.2.

The bootstrap simulations of $\hat{\theta}_b^*$ gave the following results

average	22608
standard deviation	7539
bias estimate	$22608 - 12900 = 9708.$

These values are of course estimates only. The estimate $\hat{\theta}^*$ is also judged to have a distribution with approximately the same shape as the histogram. Finally, since the bias is estimated to be very large, we compute a bias corrected estimate (for the original problem and therefore without a star)

$$\hat{\theta}_2 = 21891 - 9708 = 12183.$$

Comment
This particular bootstrap problem is somewhat special. In fact there is no need for simulations, since we can compute the results of infinitely many simulations theoretically. The estimate $\hat{\theta}_b^*$ is an average of 11 independent bootstrap observations, and both the expected value and the variance are easily derived.

$$E_*[\hat{\theta}_b^*] = E_*[X_1^*] = \frac{1600 + \ldots + 91400}{11} = 21891$$

is simply the mean of the original data, and

$$\text{Var}_*(\hat{\theta}_b^*) = \text{Var}_*(\bar{X}^*) = \frac{1}{11}\text{Var}(X_1^*)$$

$$= \frac{1}{11}\sum_{i=1}^{11}(x_i - 21891)^2 \cdot \frac{1}{11},$$

where x_i are the original data. This gives

$$\text{Var}_*(\hat{\theta}_b^*) = 53290000$$
$$\sigma_*(\hat{\theta}_b^*) = 7300$$
$$\text{Bias}^* = 21891 - 12900 = 8991.$$

The earlier simulation results are approximate estimates of these theoretical values. The exact computation changes our conclusions slightly, and in particular the bias adjusted estimate now becomes

$$\hat{\theta}_2 = 21891 - 8991 = 12900.$$

The reader may recognize this number. Starting with the mean, we have arrived at the median of the data as our bias adjusted estimator after an exact computation. This is an example of a more general observation. The bias adjusted estimate is given as $\hat{\theta} - E_*[\hat{\theta}^* - \tilde{\theta}]$, and whenever $\hat{\theta}^*$ is centred around $\hat{\theta}$ we will get $\tilde{\theta}$ as the bias adjusted parameter. Thus the definition $\hat{\theta} = \theta(F_n)$ is a very natural starting point under such circumstances.

Also the bootstrap analysis of the median estimation in Section 5.4.1 can be handled analytically with a little more effort. The simulated version is more instructive, however, and is therefore preferred for pedagogical reasons. The interested reader can at this stage try to find formulas replacing the simulations.

5.5 Double bootstrap

A bootstrapped and bias adjusted estimator will of course have somewhat different statistical properties than the estimator without such adjustment. We might therefore argue that the bootstrap adjusted estimator should also be bootstrapped. This means a double bootstrap. An illustration of such a double bootstrap is given by the combination of median estimation by the average (Section 5.4.2) and median estimation by the sample median (Section 5.4.1). Starting with the average $\bar{x}$, we found that the bootstrap estimated bias was exactly the difference between the mean and the median of the data (when the number of bootstrap replicates goes to infinity or when analysed theoretically). A bootstrap analysis of the median estimator can therefore be regarded as a double bootstrap of the mean and its bias correction. The last (outer) bootstrap analysis of the median may still show some bias and motivate another bias correction, which can in principle lead to another bootstrap and so on.

With the data of Section 5.4, the first bias correction reduces the estimate from 21891 (the average) to 12900 (the median), and bootstrapping the median leads to a further bias correction from 12900 to 10957. The bootstrap estimated variability also shows a smaller standard deviation for the median than for the mean, so the bias adjusted estimate is probably more stable in this particular case. The last estimator (10957) has not been bootstrapped, yet.

If the bias correction is bound to be small, compared to the standard deviation of $\hat{\theta}$, then the standard deviation of $\hat{\theta}^*$ is a useful estimator of the variability of both $\hat{\theta}$ and of the bias corrected estimator $\hat{\theta} -$ Bias*. But if the bias correction may be large, the variability of $\hat{\theta} -$ Bias* may well be quite different, and sometimes much larger, than that of $\hat{\theta}$. The double bootstrap always offers a possible estimate of this new standard deviation, but it may take much computer resourses. In a computer program, the inner loop of the double bootstrap should bootstrap and bias correct an original estimator, and the outer loop should bootstrap this already bootstrap corrected estimate. Suppose the statistical effects of this double bootstrap have to be studied by simulations. We then have a huge three-level simulation to handle. In this and similar situations, an analytical solution of the inner bootstrap will be of the greatest value.

5.5.1 Flow chart of double bootstrapped bias correction

In the following flow chart we produce both a bias corrected estimate and an estimate of the variations of this estimator. The studied parameter is written $\theta = \theta(F(x))$.

INITIATION

$x_1, \ldots, x_n$	original data
$F_n(x)$	empirical distribution
$\hat{\theta} = \hat{\theta}(x_1, \ldots, x_n)$	estimator of θ
$\tilde{\theta} = \theta(F_n(x))$	bootstrap parameter

OUTER BOOTSTRAP LOOP

iout$= 1, 2, \ldots$

$X_1^*, \ldots, X_n^*$	resample from x_i
$F_n^*(x)$	
$\hat{\theta}^* = \hat{\theta}(X_1^*, \ldots, X_n^*)$	
$\tilde{\theta}^* = \theta(F_n^*(x))$	

INNER BOOTSTRAP LOOP

jin$= 1, 2, \ldots$

$X_1^{**}, \ldots, X_n^{**}$	resample from X_i^*
$\hat{\theta}^{**} = \hat{\theta}(X_1^{**}, \ldots)$	
Bias$^{**} = \overline{\hat{\theta}^{**}} - \tilde{\theta}^*$	

end of inner loop

$\hat{\theta}_b^*(\text{iout}) = \hat{\theta}^* - \text{Bias}^{**}$	adjusted bootstrap est.
Var$(\hat{\theta}_b^*)$	
Bias$^* = \overline{\hat{\theta}^*} - \tilde{\theta}$	

end of outer loop

OUTPUT

$\hat{\theta}_b = \hat{\theta} - \text{Bias}^*$	adjusted original est.
Histogram of $\hat{\theta}_b^*$ or $\hat{\theta}_b^* - \tilde{\theta}$	
$\hat{\sigma}(\hat{\theta}_b) = \sqrt{\text{Var}(\hat{\theta}_b^*)}$	

5.6 Percentile estimation

Estimation of one of the percentiles, the median, was studied above. We now turn to a general percentile where we can use some different estimates. The first and obvious one is to use the corresponding percentile of the data as an estimate of the distribution's percentile. Unless we smooth the empirical distribution, a whole set of percentiles will be estimated by the same value, since the empirical distribution jumps at the data points. To be exact, all percentiles corresponding to probabilities in the open interval $(\frac{k-1}{n}, \frac{k}{n})$ will be estimated by $x_{(k)}$. If the percentile k/n is estimated, we have instead the problem that the estimator is arbitrarily defined somewhere on the flat part where $F_n(x) = k/n$. Usually we take the midpoint. These are irritating facts for small samples, but less important when samples are large. Often a very simple smoothing of $F_n(x)$ will solve all problems both for the estimation and for the definition of $\hat{\theta}$.

There are some different smoothing possibilities. We may for example use kernel estimation of the density, defined as

$$\hat{f}(x) = \Sigma \frac{1}{n} k(\frac{x - x_i}{d})/d.$$

Here the kernel $k(x)$ is in itself a density with mean zero, variance 1 (the standard normal can be used). The parameter d transforms the width of the kernel. A good value of d is important and sometimes this is determined by a separate cross validation as in Silverman (1986). Integrating $\hat{f}(x)$ we get a smooth distribution function as $\hat{F}(x) = \Sigma \frac{1}{n} K(\frac{x-x_i}{d})$ where $K(x) = \int_{-\infty}^{x} k(u)du$. For further refinements of kernel methods the reader may consult the rich literature in journals, e.g. Wand et al. (1991), the survey by Izenman (1991), Cline (1988), and Faraway and Jhun (1990).

A simple interpolation method can also be useful, especially since it runs very fast. It is given here and further generalized in Section 7.4.4. Let $x_{(1)}, \ldots, x_{(n)}$ be the ordered sample, and $X_{(1)}, \ldots, X_{(n)}$ the corresponding stochas-

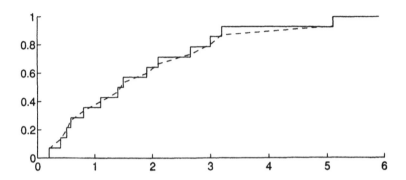

Figure 5.3 *Traditional empirical distribution and smooth version taking values $k/(n+1)$ at data points and interpolated between.*

tic variables. It can then be shown that $E[F(X_{(k)})] = \frac{k}{n+1}$. It is natural to estimate the distribution function at the observations by the expected value. In Figure 5.3 a smooth empirical distribution is based on these values, and linear interpolation between the points $(x_{(k)}, \frac{k}{n+1})$. This smooth distribution is typically left undefined outside the range of the data, which is often more reasonable than guessing the tails. For non-negative data we may sometimes extend the definition by interpolating from $(0,0)$ to $(x_{(1)}, \frac{1}{n+1})$.

The estimated percentiles and the empirical parameter $\tilde{\theta}$ can now be defined by the smooth empirical distribution, even if for simplicity we should prefer to use the traditional empirical distribution (or the original sample itself) as the basis for the bootstrap samples.

Now suppose we are willing to guess a parametric model for our data. This gives new possible estimates of the percentiles. We can estimate the model's parameters by the maximum likelihood method, or by some other estimation method, and then search the percentile in the fitted distribution. For all the models we usually apply, this will give an estimator $\widehat{\theta}_p(x_1, \ldots, x_n)$ which is a smooth function of both the percentile probability p and the data. The reader can now check his or her understanding by suggesting how $\tilde{\theta}$

should be defined in this case. There is a good analogy with the median case above.

Wrong parametric models

In many parametric models we have a good theory for how ML-estimation works. But what happens when we pick the wrong model? The theory will of course break down, but the bootstrap analysis does not care. It only uses the estimation formula from the theory and then gives us a picture of how this estimator works. (There are some versions of the bootstrap which are based on parametric assumptions and therefore are sensitive to correct model specifications. Here we are still discussing the original bootstrap.) Of course, a better model would probably have given a better estimation formula, but if we have tried our best already, and come out wrong, it is a remedy that at least we have a useful evaluation method for our estimator.

Estimating very extreme percentiles, for probabilities close to zero or one, is sometimes highly desirable. This leads into situations where we do not expect any data in the neighbourhood of the estimated parameter. Usually it will be impossible to define a useful $\tilde{\theta}$-value. Without $\tilde{\theta}$, the bootstrap can still take a suggested estimator and show how sensitive it is to data variations, but we cannot compare it to some target value.

5.7 Confidence intervals

The bootstrap analysis produces a bootstrap distribution for $\hat{\theta}^*$ or $\hat{\theta}^* - \tilde{\theta}$. It is tempting to use this distribution in a more exact way and try to produce confidence intervals for θ. How this should be done is a debated and still somewhat controversial business. When the number n of observations grows, asymptotic results have been proved for some estimators (means, percentiles, ...) saying that under some conditions

$$P^*(\sqrt{n}(\hat{\theta}^* - \tilde{\theta}) \leq x) - P(\sqrt{n}(\hat{\theta} - \theta) \leq x) \to 0, \qquad (5.12)$$

see Singh (1981). Bickel and Freedman (1981) state similar results for means, studentized means and empirical distribution functions. In their formulation both probabilities in (5.12) are conditional on $x_1, \ldots, x_n$ for every n and the convergence to zero occurs for almost every sequence $x_1, x_2, \ldots$. (Almost every means that the set of exceptional sequences has probability zero of occurring.) We will motivate our basic confidence intervals by these asymptotic results. The theory will be discussed further in Chapter 6.

5.7.1 Simple intervals

Let $X_1, \ldots, X_n$ be independent with distribution function $F(x)$. Let $\hat{\theta} = \hat{\theta}(X_1, \ldots, X_n)$ be an estimator of a parameter $\theta = \theta(F)$ which is of interest. Define $\tilde{\theta} = \theta(F_n)$ as the parameter of the empirical distribution in the usual way. Let $x_1, \ldots, x_n$ be the observed values of $X_1, \ldots, X_n$.

Resample $X_1^*, \ldots, X_n^*$ from $F_n(x)$ (or equivalently from $x_1, \ldots, x_n$). From this bootstrap sample we compute the estimator $\hat{\theta}^* = \hat{\theta}(X_1^*, \ldots, X_n^*)$ and this is simulated a large number of times, say 1000 times or more. Select a reasonable confidence level c, and find values a^* and b^* in the bootstrap distribution such that

$$P^*(a^* < \hat{\theta}^* - \tilde{\theta} < b^*) \approx c. \qquad (5.13)$$

Using the analogy (5.12), we conclude that also

$$P(a^* < \hat{\theta} - \theta < b^*) \approx c.$$

Solving for θ, our confidence interval therefore becomes

$$\hat{\theta} - b^* < \theta < \hat{\theta} - a^*, \qquad (5.14)$$

with the approximate confidence level c.

Before we give some illustrations, let us discuss how approximate we can allow such results to be. In complicated situations where no other method is available, we must be satisfied with any measure of uncertainty which has the right order of magnitude. If a nominal confidence level of 0.95 turns out to be around 0.90 the result is still satisfactory for many purposes. If the level falls below 0.80 the analysis starts to be misleading. In simpler situations, where we have parametric alternatives of a traditional kind, we usually require much more exactness. It is fair to say at once that confidence levels as high as 0.99 or more will usually not be well approximated unless the samples are very large.

5.7.2 Studentized intervals

Sometimes we can produce an estimator $\hat{\sigma}(\hat{\theta})$ of the standard deviation of $\hat{\theta}$. There is now much experience showing that the studentized expression $(\hat{\theta} - \theta)/\hat{\sigma}(\hat{\theta})$ is usually more stable than the difference $\hat{\theta} - \theta$. By more stable we mean that the probability distribution will vary less when we vary the distribution of the data. Since the empirical distribution is at some distance from the true distribution, such stability is highly desirable. We can therefore expect better confidence intervals if we bootstrap $(\hat{\theta}^* - \tilde{\theta})/\hat{\sigma}(\hat{\theta}^*)$ and translate the distribution to $(\hat{\theta} - \theta)/\hat{\sigma}(\hat{\theta})$.

In Chapter 6 we discuss some other stabilizing methods constructed by Beran (1987) and Efron (1979, 1982, 1985, and 1987). See also Hall (1988), DiCiccio and Romano (1988), and Hinkley (1988).

Example 5.2 Interval for survival probability
In a large installation of electric bulbs, all the bulbs are planned to be replaced regularly after 1200 hours. In order to form an opinion about this strategy, the probability that a bulb will survive that many hours is of interest. We will estimate this probability by a confidence interval. A limited test is run and gives the following 20 observed life times.

Data: x_i, $1 \le i \le 20$.

1354	1552	1766	1325	2183	1354	1299
627	695	2586	2420	71	2195	1825
159	1577	3725	884	1014	965	

Our model is that X_i are independent and identically distributed with a continuous distribution function $F(x)$. Introduce the parameter

$$\theta = 1 - F(1200),$$

which is the probability of interest. Suppose the estimate

$$\hat{\theta} = e^{-1200/\bar{x}} = 0.444$$

is used. This estimate can be motivated if X_i are exponentially distributed, since in this distribution $\theta = e^{-1200/\mu}$,
where $\mu = E[X_i]$. Based on this estimator we will construct
a 90% bootstrap confidence interval for θ. According to the
bootstrap ideas, this construction should be valid even if the
distribution is somewhat different from the exponential.

The empirical distribution $F_n(x)$ has the parameter

$$\tilde{\theta} = 1 - F_n(1200) = 13/20,$$

since thirteen of the observations are bigger than 1200.

Draw a bootstrap sample $X_1^*, \ldots, X_{20}^*$ from the given data.
From this sample we get

$$\hat{\theta}^* = e^{-1200/\bar{X}^*}.$$

The results of 1000 such bootstrap estimates are shown as
a histogram in Figure 5.4. From this simulation we find
that 90% of the bootstrap replicates fall between 0.356 and
0.514. We have then excluded 5% in each tail. Subtracting
$\tilde{\theta}$ from all the results, all bootstrap differences $\hat{\theta}^* - \tilde{\theta}$ turn
out to be negative, and we get $a^* = 0.356 - \tilde{\theta} = -0.294$, and
$b^* = 0.514 - \tilde{\theta} = -0.135$,

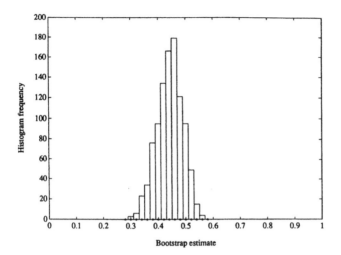

Figure 5.4 *Histogram of 1000 bootstrap replicates of* $\exp(-1200/\bar{X}^*)$.

$$P^*(a^* < \hat{\theta}^* - \tilde{\theta} < b^*) \approx 0.90.$$

The same conclusion for $\hat{\theta} - \theta$ becomes

$$P(a^* < \hat{\theta} - \theta < b^*) \approx 0.90,$$

and solving the inequality for θ, we arrive at the confidence interval $0.444 + 0.135 < \theta < 0.444 + 0.294$ or

$$0.579 < \theta < 0.738$$

with approximately 90% confidence level.

A studentized version of the confidence interval construction may be designed as follows. From the original data we compute $\bar{x} = 1479$, $s = 869.8$, $s/\sqrt{n} = 194.4$ and $n = 20$. For the estimate $\hat{\theta} = \exp(-1200/\bar{x})$ we can approximate the standard deviation by linear approximation if variations in $\bar{x}$ are regarded as small perturbations. We then have

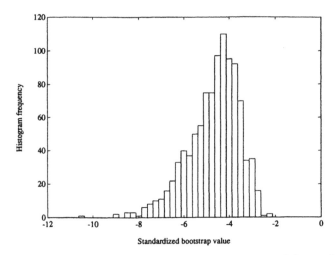

Figure 5.5 *Histogram of 1000 bootstrap replicates of the studentized variable.*

$$\sigma_{\hat{\theta}} \approx |\frac{d\hat{\theta}}{d\bar{x}}|\sigma_{\bar{x}},$$

and estimate this as

$$\hat{\sigma}_\theta = e^{-\frac{1200}{\bar{x}}}\frac{1200}{\bar{x}^2}s/\sqrt{n} = 0.0474 \ .$$

The expression $R = (\hat{\theta} - \theta)/\hat{\sigma}_\theta$ can now be bootstrapped. From each bootstrap sample we compute $\bar{x}^*$ and s^* and find $\hat{\sigma}_\theta^*$ by the same expression as for $\hat{\sigma}_\theta$ above. The histogram of 1000 simulated $R^* = (\hat{\theta}^* - \tilde{\theta})/\hat{\sigma}_\theta^*$ is shown in Figure 5.5. The central 90% of the values fall between -6.801 and -3.106, indicating again the strong bias of $\hat{\theta}$. We therefore state with approximately 90% confidence that

$$-6.801 < \frac{\hat{\theta} - \theta}{\hat{\sigma}_\theta} < -3.106, \quad (90\%)$$

or $0.591 < \theta < 0.767$.

Remark 1

That all bootstrap differences are negative illustrates that
the exponential model is probably very poor for this kind
of electric bulb. Without any such parametric model we can
use the binomial distribution to estimate θ, since the number
Y of surviving units is Bi$(20, \theta)$. Observing 13 such units out
of 20, a 90% confidence interval is roughly $0.45 < \theta < 0.82$.

The given data were drawn at random from a larger sample
of 120 observations. Of these, the proportion 0.63 survived
1200 hours. Another use of the binomial distribution gives
that the true θ should be within 0.63 ± 0.09 with 95% confidence. The bootstrap interval is in good agreement with this
result, but we have no possibility at this stage to judge how
close to 90% the true level of confidence is for the bootstrap
interval.

Remark 2

Theoretical results concerning studentized and other refined
confidence intervals are discussed further in Chapter 6. The
higher order of convergence for such methods is based on
some assumptions like $\theta = \theta(F)$, a function of the distribution, and $\hat{\theta} = \theta(F_n)$, the same function of the estimated distribution. Then $\tilde{\theta} = \hat{\theta}$ is natural. Our situation violates this
theory. The estimate $\exp(-1200/\bar{x})$ is based on the exponential distribution. It is a function of F_n, since $\bar{x}$ is so, but
if the exponential distribution is wrong, as the data strongly
indicate, the true probability θ is not the same function of
F. Assuming this, $\hat{\theta} - \theta$ will typically not converge to zero
but to some constant b and we have

$$R = \frac{\hat{\theta} - \theta}{\hat{\sigma}_\theta} = \frac{b}{\hat{\sigma}_\theta} + \frac{\hat{\theta} - \theta - b}{\hat{\sigma}_\theta}.$$

The statistical behaviour of $b/\hat{\sigma}_\theta$ can easily dominate that
of the almost unbiased $(\hat{\theta} - \theta - b)/\hat{\sigma}_\theta$ and the asymptotic
theory is relevant for the last term only.

5.8 Statistical small sample properties

It is now time to discuss the statistical properties of boot-strap methods. So far we have worked under the assumption that the methods are sound and give reasonable measures. It would be nice to have a proof that estimated measures of variability, e.g. standard deviations, are good, that bias adjusted estimates are almost unbiased or at least better than estimates without such adjustments, and that bootstrap confidence intervals have approximately the correct degree of confidence. Unfortunately there is no general proof available for finite samples. There are, however, results of some generality for the asymptotic situation when sample sizes go to infinity. One alternative approach for the small sample situation is then to look at special cases and study how the bootstrap works. In particular we can study classical situations and compare with well-known parametric solutions. By simulation methods we can also study some situations which are not so easily analysed by classical methods.

Example 5.3   Estimation of probability
Let $X_1, \ldots, X_n$ be independent. Each $X_i = 1$ with probability p and $X_i = 0$ otherwise. The classical estimation of p goes as follows.

$$\hat{p} = \frac{\sum X_i}{n} = \bar{X}$$

$$E\hat{p} = p$$

$$\text{Var}(\hat{p}) = \frac{p(1-p)}{n}$$

$$\widehat{\text{Var}}(\hat{p}) = \frac{\hat{p}(1-\hat{p})}{n}.$$

We find that $\hat{p}$ is unbiased and estimate its variance as in the last row.

The bootstrapper resamples $X_1^*, \ldots, X_n^*$ by drawing each value at random from the observed values $x_1, \ldots, x_n$ of the variables. Conditional on the original data the X_i^* are independent with distribution

$$X_i^* = \begin{cases} 1 & \text{with probability } \hat{p} \\ 0 & \text{with probability } 1 - \hat{p}, \end{cases} \qquad (5.15a)$$

since $\hat{p}$ is the proportion of ones in the original sample. The corresponding empirical distribution function becomes

$$F_n(x) = \begin{cases} 0 & \text{if } x < 0 \\ 1 - \hat{p} & \text{if } 0 \le x < 1 \\ 1 & \text{if } x \ge 1. \end{cases} \qquad (5.15b)$$

The parameter of the empirical distribution (5.15a or b) is also given by the proportion of ones in the original sample and therefore becomes $\tilde{p} = \hat{p}$. Let the bootstrap estimator be

$$\hat{p}^* = \frac{\sum X_i^*}{n}.$$

This estimator gets

$$E_* \hat{p}^* = \hat{p}, \qquad \text{Var}_*(\hat{p}^*) = \frac{\hat{p}(1 - \hat{p})}{n}. \qquad (5.16)$$

These results can of course be found approximately by simulations, but are also known analytically since we have an ordinary binomial distribution in the bootstrap step.

The first conclusions of our analysis is that $E_* \hat{p}^* = \tilde{p}$ or Bias$^* = 0$, so our estimator seems unbiased which is correct. The second conclusion is that $\hat{p}$ has the bootstrap variance. This variance is the same as the variance we estimate by classical methods. We have therefore got quite reasonable results for the probability estimation.

Example 5.4   The mean
Let $X_1, \ldots, X_n$ be a sample of independent and identically distributed variables with expected value μ and variance σ^2. We will study the estimate $\hat{\mu} = \bar{X}$. Theory gives that

$$E\hat{\mu} = \mu, \qquad \text{Var}(\hat{\mu}) = \sigma^2/n.$$

The bootstrap method means that given the observed sample $x_1, \ldots, x_n$ and its empirical distribution $F_n(x)$, the bootstrap sample $X_1^*, \ldots, X_n^*$ is drawn from $F_n(x)$ or equivalently from the values $x_1, \ldots, x_n$ in the original sample. Every X_i^* has therefore

$$E_* X_i^* = \bar{x} = \hat{\mu}$$

and

$$\text{Var}_*(X_i^*) = \sum_{i=1}^{n} \frac{1}{n}(x_i - \bar{x})^2 = \frac{n-1}{n} s^2.$$

This means that the empirical distribution $F_n(x)$ has the expected value $\tilde{\mu} = \hat{\mu}$ and the variance $\tilde{\sigma}^2 = ((n-1)/n)s^2$. The bootstrap estimate $\hat{\mu}^* = \overline{X^*}$ therefore gets

$$E_* \hat{\mu}^* = E_* \overline{X^*} = \hat{\mu}$$
$$\text{Bias}^* = 0 \qquad\qquad\qquad (5.17)$$
$$\text{Var}(\hat{\mu}^*) = \frac{1}{n}\text{Var}(X_i^*) = \frac{n-1}{n^2} s^2.$$

This leads to the correct conclusion that $\hat{\mu}$ is unbiased since $\hat{\mu}^*$ is so. We also find that the variance of $\hat{\mu}$ will be estimated by $(n-1)s^2/n^2$ instead of the ordinary estimate s^2/n of the theoretical value σ^2/n. The variance is therefore underestimated, but not much unless n is very small. The bootstrap estimate of the variance is in fact the same as the maximum likelihood estimate for Gaussian variables. In summary we can say that the bootstrap works well for the mean.

The result for the mean can be extended to all moments $\mu_k = E[X^k]$, since $Y = X^k$ can be seen as a new variable. In the exceptional case when the expected value or the variance of X^k does not exist or is infinite, the bootstrap estimates will of course be misleadingly finite, if all observations are finite. One such example is the Cauchy distribution with $f(x) = \text{constant}/(1 + x^2)$, where not even the first moment exists. It is hardly a serious objection that the bootstrap fails in such cases.

5.9 Bootstrap as a definer of functions

The distribution of a bootstrap sample is given by the empirical distribution $F_n(x)$. This in turn is a function of the original data $x_1, \ldots, x_n$. Every theoretical quantity based on the bootstrap probability measure P^* will be conditional on the distribution F_n. It is therefore a function of F_n or more explicitly of $x_1, \ldots, x_n$. This refers to $\tilde{\theta}$, $P^*(\hat{\theta}^* - \tilde{\theta} \le x)$, $\text{Bias}^* = E_*[\hat{\theta}^* - \tilde{\theta}]$ and so on. If we use bootstrap simulations, we aim at such theoretical quantities and get them with some error due to the randomness (noise) of the simulations. With large enough simulations, this noise will be unimportant. We can then regard the simulated results also as functions of the original data. It is in fact clarifying to see the bootstrap as a way to define complicated and useful functions of a sample.

Let $\hat{\theta}_2 = \hat{\theta} - \text{Bias}^*$ be a bias adjusted estimate of θ, and let a confidence interval $a < \theta < b$ also be based on the bootstrap analysis. We consider the basic structure of the problem and the bootstrap analysis as given in a well defined way. Then the following box description applies

$$\begin{bmatrix} \text{Input} \\ x_1, \ldots, x_n \\ \hat{\theta}(x_1, \ldots, x_n) \end{bmatrix} \longrightarrow \begin{bmatrix} \text{Bootstrap} \\ \text{analysis} \\ P^*,\ \tilde{\theta},\ E_* \\ \text{Bias}^* \ldots \end{bmatrix} \longrightarrow \begin{cases} \hat{\theta}_2(x_1, \ldots, x_n) \\ a(x_1, \ldots, x_n) \\ b(x_1, \ldots, x_n) \end{cases}$$

where the functions $\hat{\theta}_2(.)$, $a(.)$, $b(.)$, are defined by the bootstrap analysis, but in the end only depend on the input data (plus noise).

Every classical statistical inference method will also produce functions of the data, and there is in fact no logical difference between bootstrap functions and more classical functions. The statistical properties of the results have to be studied in the same manner.

The interesting thing is therefore to study $\hat{\theta}_2(X_1, \ldots, X_n)$ and the degree of confidence for the interval, etc., with the probability distribution valid for the original data $x_1, \ldots, x_n$. This requires that we know enough about the distribution of the data to be able to carry through the analysis. Many estimation problems can therefore only be studied in special cases using a chosen distribution $F_0(x)$ for the original data. If we want to simulate to get the properties of the bootstrap results, we have to generate observations $x_1, \ldots, x_n$ a large number of times. For each such set of simulated observations, we perform the bootstrap analysis typically by making new inner simulations with bootstrap data $X_1^*, \ldots, X_n^*$ drawn from the generated values $x_1, \ldots, x_n$. The bootstrap analysis will produce results like $\hat{\theta}_2$, a, b, etc. The confidence level of the interval $a < \theta < b$ can now be measured as the number of times the true parameter of $F_0(x)$ will be hit by the interval (a, b) divided by the number of generated samples $(x_1, \ldots, x_n)$. In a similar way all other results of the bootstrap analysis can be studied. It is obvious that such simulation studies of an already computer intensive method can be very computer intensive and may easily pass the border of a computer's capacity. One of the many such studies is illustrated in our next example.

Example 5.5   Correlation uncertainty
The following example is part of a simulation study described in Efron (1982. p. 18). The problem formulation has been slightly adapted for our text.

Original problem: We have 14 two-dimensional observations $x_1, \ldots, x_n$, $n = 14$, and estimate the correlation ρ in the usual way. What standard deviation has the estimate?

Simulated problem: Let $X_1, \ldots, X_n$ be independent vectors which are two-dimensional normal with the correlation $\rho = 0.5$. The means and variances of the variables will cancel out, so we may arbitrarily use mean zero and variance one for the vector components. Simulate 200 such vectors of observations. Bootstrap simulate 512 times for each vector of observations and compute $\sigma_*(\hat{\rho}^*)$. Efron also performed a jackknife estimation of the standard deviation for each sample $x_1, \ldots, x_n$ as comparison. We will not go into the details of this method here. The following statistics illustrate how good the estimated standard deviations are

	Average of $\sigma_*(\hat{\rho}^*)$	Standard deviation
Bootstrap	0.206	0.063
Jackknife	0.223	0.085
True value	0.218	

(Reproduced with the kind permission of the author).

The amount of computation behind this little table is 200 outer simulations $\times$ 512 bootstrap replicates $\times$ 14 data $\times$ 2 dimensions = 2 867 200 generated observations and 200 $\times$ (512+1) = 102 600 estimates of correlations. Later we will come to considerably larger computations, but this already gives an idea of the magnitudes.

Example 5.6   Median estimation
We return to our introductory example, where the median of the distribution was estimated by the sample median. Again we use eleven observations. This time we add that the sample is from the exponential distribution, but this information is not used in the bootstrap analysis.

Let $X_1, \ldots, X_{11}$ be independent and exponentially distributed. Let θ be the median of the distribution and let

$$\hat{\theta} = X_{(6)},$$

where $X_{(1)} \leq X_{(2)} \leq \ldots \leq X_{(11)}$ is the ordered sample. In our computations below we will use the Exp(1)-distribution with density $f(x) = e^{-x}$, $x > 0$. This is no limitation since the expected value μ is a pure scale parameter. Multiplication of all X_i with μ gives the Exp(μ)-distribution and the results are easily translated.

The theoretical distribution of $\hat{\theta} = X_{(6)}$ is given by the density

$$f_{\hat{\theta}}(x) = 11\binom{10}{5}(1 - e^{-x})^5 e^{-6x}, \quad x > 0,$$

but we can also imagine a (time-)axis where the values $X_{(1)}$, $X_{(2)}, \ldots, X_{(11)}$ are plotted. It can then be shown that the differences

$$X_{(1)}, X_{(2)} - X_{(1)}, \ldots, X_{(11)} - X_{(10)}$$

are independent variables which are exponential with intensities 11, 10, 9,..., 1 and therefore have expected values $1/11, 1/10, \ldots, 1$.

(Suppose 11 units with independent exponential life times are started at time zero. The time until the first breaks, $X_{(1)}$, is then exponential and gets 11 times the individual intensity. The residual life time of each surviving unit is again exponential with the same individual parameter as before, since the exponential distribution is memoryless and has $P(X > t + s | X > t) = P(X > s)$. The time to the next failure, $X_{(2)} - X_{(1)}$, is therefore exponential with ten times the individual intensity and so on.)

Using this we easily get for $\hat{\theta} = X_{(6)}$

$$E[\hat{\theta}] = \frac{1}{11} + \frac{1}{10} + \frac{1}{9} + \frac{1}{8} + \frac{1}{7} + \frac{1}{6} = 0.7365$$

$$\mathrm{Var}(\hat{\theta}) = \frac{1}{121} + \frac{1}{100} + \frac{1}{81} + \frac{1}{64} + \frac{1}{49} + \frac{1}{36} = 0.0944$$

$$\sigma_{\hat{\theta}} = 0.307.$$

The true median θ in Exp(1) is given by $F(\theta) = 1 - e^{-\theta} = 1/2$.

$$\theta = \ln 2 = 0.6931.$$

We therefore know that our estimator has the wrong expected value, Bias $= 0.7365 - 0.6931 = 0.0434$, and the variance 0.0944. We are now ready to study the bootstrap method.

Bootstrap analysis
Draw a bootstrap sample $X_1^*, \ldots, X_{11}^*$ at random and independently from the observed values $x_{(1)}, \ldots, x_{(11)}$. The corresponding empirical distribution F_n has the median $\tilde{\theta} = x_{(6)}$. Let $X_{(1)}^*, \ldots, X_{(11)}^*$ be the ordered bootstrap sample. The median $X_{(6)}^*$ of the bootstrap sample must be one of the values $x_{(1)}, \ldots, x_{(11)}$. If six or more of the X_i^* are at most $x_{(k)}$, then the median $\hat{\theta}^*$ will be at most $x_{(k)}$. Define $a_0 = 0$ and introduce for $1 \leq k \leq 11$,

$$a_k = P_*(X_{(6)}^* \leq x_{(k)}) = \sum_{r=6}^{11} \binom{11}{r} \left(\frac{k}{11}\right)^r \left(1 - \frac{k}{11}\right)^{11-r}$$

$$p_k = a_k - a_{k-1} = P_*(X_{(6)}^* = x_{(k)}).$$

(5.18)

The values p_k are computed in Table 5.2.

These probabilities are of course independent of the $x_{(i)}$-values. If we simulate long enough in the bootstrap analysis for a given set of data $x_1, \ldots, x_{11}$ we will therefore find that

$$E_* \hat{\theta}^* = \sum p_k x_{(k)}$$

$$\text{Bias}_* = \sum p_k x_{(k)} - x_{(6)}$$

$$\hat{\theta}_2 = x_{(6)} - \text{Bias}_* = 2x_{(6)} - \sum p_k x_{(k)} \qquad (5.19)$$

$$\text{Var}_*(\hat{\theta}^*) = \sum p_k (x_{(k)} - E_* \hat{\theta}^*)^2$$

$$= \sum p_k x_{(k)}^2 - \left(\sum p_i x_{(i)}\right)^2.$$

We can now check whether the adjusted estimator is unbiased or not, and if $\text{Var}(\hat{\theta}^*)$ on average gives $\text{Var}(\hat{\theta})$, and also compare $\mathrm{Var}(\hat{\theta}_2)$ with $\mathrm{Var}(\hat{\theta})$. Answering these questions requires that we take expected values with respect to the variations of the original sample $x_1, \ldots, x_{11}$. This corresponds to the previously mentioned outer simulation in pure simulation studies. We want the expected values

$$E\hat{\theta}_2 = 2E[X_{(6)}] - \sum p_k E[X_{(k)}]$$
$$E[\mathrm{Var}_*(\hat{\theta}^*)] = \sum p_k E[X_{(k)}^2] - \sum \sum p_i p_j E[X_{(i)} X_{(j)}].$$

Let $\mu_k = EX_{(k)}$ and $v_k = \mathrm{Var}(X_{(k)})$. The earlier representation of $X_{(k)}$ as a sum of independent exponentially distributed pieces $X_{(1)}, X_{(2)} - X_{(1)}, \ldots, X_{(k)} - X_{(k-1)}$ gives easily that

$$\mu_k = \sum_{r=0}^{k-1} \frac{1}{11 - r}, \qquad v_k = \sum_{r=0}^{k-1} \frac{1}{(11 - r)^2}, \qquad (5.20)$$

where we have used the mean and variance of the exponential distribution. The numerical values are given in Table 5.2.

Table 5.2. Probabilities (5.18) and moments (5.20).

k	p_k	μ_k	v_k
1	0.000174	0.0909	0.0083
2	0.00703	0.1909	0.0183
3	0.0440	0.3020	0.0306
4	0.1215	0.4270	0.0462
5	0.2059	0.5699	0.0666
6	0.2427	0.7365	0.0944
7	0.2059	0.9365	0.1344
8	0.1215	1.1865	0.1969
9	0.0440	1.5199	0.3080
10	0.00703	2.0199	0.5580
11	0.000174	3.0199	1.5580

We also have for $i \leq j$, due to independence,

$$
\begin{aligned}
E[X_{(i)}X_{(j)}] &= E[X_{(i)}(X_{(i)} + X_{(j)} - X_{(i)})] \\
&= E[X_{(i)}^2] + E[X_{(i)}]E[X_{(j)} - X_{(i)}] \\
&= v_i + \mu_i^2 + \mu_i(\mu_j - \mu_i) = v_i + \mu_i\mu_j.
\end{aligned}
$$

Using these expressions and our earlier values p_k we find that

$$E[\hat{\theta}_2] = 0.6918$$
$$E[\mathrm{Var}_*(\hat{\theta}^*)] = 0.1342.$$

The estimator $\hat{\theta}_2$ is close to unbiased, (true $\theta = 0.6931$) so the bias correction has been quite successful. The variance estimator is on average higher than the true variance of $\hat{\theta}$ (0.1342 against $v_6 = 0.0944$). In the same way we also find that the bias adjusted estimator $\hat{\theta}_2 = 2X_{(6)} - \sum p_k X_{(k)}$ will in turn get

$$\mathrm{Var}(\hat{\theta}_2) = 0.1368$$

which is substantially higher than the variance v_6. We have removed most of the bias, but at the expense of an increased variance in the exponential case. The close agreement between $E[\mathrm{Var}_*(\hat{\theta}^*)]$ and $\mathrm{Var}(\hat{\theta}_2)$ is a pure coincidence and without interest since we have aimed at an imitation of $\hat{\theta}$ and not $\hat{\theta}_2$ in the bootstrap analysis.

The results we have found for 11 observations from an exponential distribution may have some interest in themselves. The main message is however to illustrate how the bootstrap results become functions of the original data, and in the end it is their statistical properties with respect to the probability measure for the original data that really matters.

Example 5.7 Confidence intervals for probability

We will compare two different confidence intervals for the probability p. Both intervals are based on the same estimator $\hat{p}$ given by the observed relative frequency of events. One interval studies the difference $\hat{p} - p$ and the other uses a studentized difference. The example illustrates both confidence interval constructions, how small sample properties are investigated, and tabular functions defined by the bootstrap.

Interval 1

Let $Y = X_1 + \ldots + X_n$ be binomial $\mathrm{Bi}(n,p)$, with X_i as in Example 5.3, and estimate p by $\hat{p} = Y/n$. Given $Y = y$, let $Y^* = X_1^* + \ldots + X_n^*$ be binomial $\mathrm{Bi}(n,\tilde{p})$ where $\tilde{p} = \hat{p} = y/n$. Define $\hat{p}^* = Y^*/n$.

For each possible value of $\tilde{p}$ we study the exact probabilities of the binomial distribution for Y^*. This replaces a very long bootstrap simulation. We then find a maximal $a^* = a^*(\tilde{p})$ and a minimal $b^* = b^*(\tilde{p})$, both functions of $\tilde{p}$, such that $P_*(a^* \leq \hat{p}^* - \tilde{p}) \geq 0.95$, $P_*(\hat{p}^* - \tilde{p} \leq b^*) \geq 0.95$, and therefore also $P_*(a^* \leq \hat{p}^* - \tilde{p} \leq b^*) \geq 0.90$. These inequalities are now translated into the new statement $a^* \leq \hat{p} - p \leq b^*$ with approximately 90% confidence level.

In Table 5.3 we illustrate a case with $n = 10$. Each column gives the cumulative distribution for Y^* conditional on the value y of Y, or equivalently given $\tilde{p}$. The positions marked by boldface numbers embrace 90% or more of the probability. After subtraction of $\tilde{p}$ from r_{bold}/n, the row numbers of the boldface positions divided by n, we have the limits a^* and b^*. When $y = 0$ or n, let $a^* = b^* = 0$.

For any given probability p, we may now check for which integers y, $0 \leq y \leq n$, this construction will contain the true p-value. Using the binomial probabilities for Y, we get the true confidence level as

$$c1 = P\left(\frac{Y}{n} - b^*(Y/n) \leq p \leq \frac{Y}{n} - a^*(Y/n)\right). \quad (5.21)$$

Table 5.3 $1000\,P(Y^* \leq r|Y = y)$ and a^*, b^* for columns (y) 1 to 9 and rows (r) 0 to 10, $n = 10$.

$y = 1$	2	3	4	5	6	7	8	9
$r=0$ 349	107	28	6	1	0	0	0	0
1 736	376	149	46	11	2	0	0	0
2 930	678	383	167	55	12	2	0	0
3 987	879	650	382	172	55	11	1	0
4 998	967	850	633	377	166	47	6	0
5 1000	994	953	834	623	367	150	33	2
6 1000	999	989	945	828	618	350	121	13
7 1000	1000	998	988	945	833	617	322	70
8 1000	1000	1000	998	989	954	851	624	264
9 1000	1000	1000	1000	999	994	972	893	651
10 1000	1000	1000	1000	1000	1000	1000	1000	1000
a^* -0.1	-0.2	-0.2	-0.2	-0.3	-0.3	-0.2	-0.2	-0.2
b^* 0.2	0.2	0.2	0.3	0.3	0.2	0.2	0.2	0.1

Interval 2

The situation is the same but we will use an estimated standard deviation and studentize before we search the confidence limits. Since $\hat{p}$ has the estimated variance $\hat{p}(1 - \hat{p})/n$, we will construct a bootstrap interval for

$$\frac{\hat{p} - p}{\sqrt{\hat{p}(1 - \hat{p})/n}}.$$

We also have the possible alternative, which is probably superior, to use $\sqrt{p(1 - p)/n}$ in the binomial situation, but using an estimate is more typical for bootstrap situations in general. For each possible value of $\tilde{p}$ we use the same binomial distribution for Y^* as before and exclude the same states as in Table 5.3. We then find a maximal $a2^*$ and a minimal $b2^*$, both functions of $\tilde{p}$ or equivalently of Y, such that

Table 5.4 *Achieved confidence*
levels for n = 5 to 70, step 5,
p = 0.3.

n	$c1$	$c2$
5	0.80	0.83
10	0.80	0.93
15	0.82	0.95
20	0.88	0.98
25	0.87	0.93
30	0.84	0.91
35	0.90	0.90
40	0.88	0.91
45	0.85	0.90
50	0.90	0.87
55	0.89	0.92
60	0.88	0.91
65	0.88	0.92
70	0.90	0.88

$$P_*\left(a2^* \le \frac{\hat{p}^* - \tilde{p}}{\sqrt{\hat{p}^*(1 - \hat{p}^*)/n}}\right) \ge 0.95$$

and

$$P_*\left(\frac{\hat{p}^* - \tilde{p}}{\sqrt{\hat{p}^*(1 - \hat{p}^*)/n}} \le b2^*\right) \ge 0.95.$$

When $\tilde{p} = 0$ or 1, $\hat{p}^*$ will be 0 or 1 as well. Define $a2^*(0) = b2^*(0) = 0$ and $a2^*(1) = b2^*(1) = 1$. The confidence statement $a2^* \le (\hat{p} - p)/\sqrt{\hat{p}(1 - \hat{p})/n} \le b2^*$ follows. With $\hat{p} = Y/n$ we arrive at the true confidence level

$$c2 = P\left(\frac{Y}{n} - b2^*(Y/n)\sqrt{\frac{\frac{Y}{n}(1 - \frac{Y}{n})}{n}} \leq p\right.$$
$$\left. \leq \frac{Y}{n} - a2^*(Y/n)\sqrt{\frac{\frac{Y}{n}(1 - \frac{Y}{n})}{n}}\right). \tag{5.22}$$

The two bootstrap intervals are both pure functions of Y. The bootstrap variable Y^* has only acted as a definer of, and motivation for, the functions $a2^*$, $b2^*$, a^*, b^*, and after that its role is over.

For $p = 0.3$ we get the confidence levels of Table 5.4 for various sample sizes n. As is easily seen from Table 5.3, we sometimes get much more than 90% in the conditional bounds for Y^* given Y. This happens especially for small n. Nevertheless the true confidence level may fall well below 90% in some cases.

The general impression is that the studentized interval is more reliable for small n. That $c2$ exceeds 90% is not surprising since 90% probability will not be achieved exactly in the binomial distributions, and we have then defined our intervals in a conservative manner.

5.10   Example of a bootstrap program, FORTRAN code

```
C       Bootstrap program developed for teaching
C       Mathematical Statistics, Linkoping University, UH

        DIMENSION X(100),XB(100),PBO(400),IHIST(20),Y(100)
        CHARACTER*30 DATFIL
        WRITE(*,*)'Bootstrap analysis of percentile, estimated with'
        WRITE(*,*)'the Weibull distribution'
        WRITE(*,*)'Limitations: max 100 data, 400 replicates.'
        WRITE(*,*)'The datafile must start with the number of data'

        WRITE(*,*)' '
        WRITE(*,*)'Give the name of the data file'
        READ(*,99)DATFIL
99      FORMAT(A)
        OPEN(21,FILE=DATFIL)
        READ(21,*)N
        READ(21,*)(X(I),I=1,N)
C******************************************************************

        WRITE(*,*)' '
        WRITE(*,*)'Ready for estimation.'
        WRITE(*,*)' Give percentile between .05 and .95'
        READ(*,*) F
        CALL WEI(X,N,A,B)
C       computation of precentile for estimated weibull param. A,B
        PTAK=(-B*ALOG(1.-F)/A)**(1./B)
        WRITE(*,*)' '
        WRITE(*,*)' Estimated percentile = ',PTAK

        pause
C******************************************************************
        WRITE(*,*)' '
        WRITE(*,*)'Study the estimate with the bootstrap method'

C       The parameter of the empirical distribution (tilde)
C******************************************************************
        CALL ORDNA(X,N,Y)
        M=INT(F*FLOAT(N))
        IF(FLOAT(M+1)/FLOAT(N+1).LT.F)M=M+1
        PEMP=Y(M)+(F-FLOAT(M)/FLOAT(N+1))
1       *(Y(M+1)-Y(M))/(1./FLOAT(N+1))
C******************************************************************
        WRITE(*,*)
        WRITE(*,*)'The parameter's value in the empirical distr =',PEMP
        WRITE(*,*)' '
        WRITE(*,*)' '
        WRITE(*,*)' Give the number of boot-replicates, max 400'
        READ(*,*)NSIM
        DO 300 ISIM=1,NSIM
        WRITE(*,8)ISIM
8       FORMAT(I5,$)
```

```
C       Make bootstrap samples
        DO 210 J=1,N
        R=RAN(8)
        INDEX=INT(R*FLOAT(N))+1
        IF(INDEX.GT.N)INDEX=N
210     XB(J)=X(INDEX)
C*******************************************************************
        CALL WEI(XB,N,A,B)
        PBO(ISIM)=(-B*ALOG(1.-F)/A)**(1./B)

C*******************************************************************
300     CONTINUE
        WRITE(*,*)' '
        WRITE(*,*)'Analysis of bootstrap results'
        pause
        PM=0.
        PV=0.
        DO 310 I=1,NSIM
310     PM=PM+PBO(I)
        PM=PM/FLOAT(NSIM)
        DO 320 I=1,NSIM
320     PV=PV+(PBO(I)-PM)**2
        PV=PV/FLOAT(NSIM-1)
        SIG=SQRT(PV)
        WRITE(*,*)' '
        WRITE(*,*)'ììììììììììììììììììììììììììììììììììììììììììììììì'
        WRITE(*,*)' '
        WRITE(*,*)'Boot-mean                    ',PM
        WRITE(*,*)'Bias=bootmean -empirical param. ',PM-PEMP
        WRITE(*,*)'Boot-standard deviation       ',SIG
        WRITE(*,*)' '
        WRITE(*,*)'Bootstrap adjusted estimate of the original estimate'
        WRITE(*,*)'                         ',PTAK-PM+PEMP
        WRITE(*,*)' '
        WRITE(*,*)'iiiiiiiiiiiiiiiiiiiiiiiiiiiiiiiiiiiiiiiiiiiii'
        WRITE(*,*)' '
        pause
        WRITE(*,*)'Histogram of bootstrap simulations'
        V=.9E+32
        H=-.9E+32
        DO 400 I=1,NSIM
        V=AMIN1(V,PBO(I))
        H=AMAX1(H,PBO(I))
400     CONTINUE
        DO 410 I=1,NSIM
        INDEX=INT(20.*(PBO(I)-V)/(H-V))+1
        IF(INDEX.GT.20)INDEX=20
        IHIST(INDEX)=IHIST(INDEX)+1
410     CONTINUE
        WRITE(*,*)'Interval      .    Number of observations'
        DO 420 I=1,20
        C=FLOAT(I-1)*(H-V)/20.+V
```

```
        D=C+(H-V)/20.
        WRITE(*,9)C,D,IHIST(I)
420     CONTINUE
9       FORMAT(2F10.3,I10)
        pause
        WRITE(*,*)' '
        WRITE(*,*)'Smallest, largest result.',V,H
        END

C
        SUBROUTINE ORDNA(X,N,Y)
        DIMENSION X(100),Y(100)
        DO 5 I=1,N
5       Y(I)=X(I)
        DO 10 I=1,N-1
        DO 10 J=I+1,N
        IF(Y(I).LE.Y(J))GOTO 10
        S=Y(I)
        Y(I)=Y(J)
        Y(J)=S
10      CONTINUE
        RETURN
        END


        SUBROUTINE WEI(X,NX,A,B)
C       WEIBULL FITTED TO COMPLETE DATA
        DIMENSION X(100)
        DIMENSION XLOG(100)
        DO 10 I=1,NX
10      XLOG(I)=ALOG(X(I))
        B=1.
        XLOGM=0.
        DO 30 I=1,NX
30      XLOGM=XLOGM+XLOG(I)/NX
40      TB=0.
        T2B=0.
        ANB=0.
        DO 50 I=1,NX
        TB =TB +X(I)**B*XLOG(I)
        T2B=T2B+X(I)**B*XLOG(I)**2
50      ANB=ANB+X(I)**B

        U=1/B + XLOGM - TB/ANB
        UPRIM=-1/B**2 - T2B/ANB + (TB/ANB)**2
        BPREL=B-U/(UPRIM+U/(2*B))
        IF(BPREL.LT.0.) BPREL=B/3.
        B=BPREL
        IF(ABS(U).GT.MAX(1/B,ABS(XLOGM),TB/ANB)/10000)GOTO 40
        A=NX/ANB
        RETURN
        END
```

5.11 Bootstrap exercises

In order to stimulate the reader's imagination, and perhaps also start some activity on his or her computer, we will indicate some problems which can be analysed by the bootstrap technique. The range of possible uses seems to be without bounds, but sometimes there can be difficulties (and of course sometimes other methods may well be superior). If trouble arises it is often associated with the definition of an empirical distribution and its tilde parameter. It is therefore recommended to consider early on how $\tilde{\theta}$ should be interpreted.

Exercise 5.1 Give your definition of $\tilde{\theta}$ when independent identically distributed data $x_1, \ldots, x_n$ are observed and the following parameters are studied:

a) $\theta = 25\%$ percentile in a distribution.

b) $\theta = P(10 < X \le 25)$ and data 'cover' this interval.

c) $\theta = \mu/\sigma$.

d) $\theta = E[\sin X]$.

e) $\theta =$ correlation and each x_i is two-dimensional.

f) $\theta =$ covariance matrix and the x_i are m-dimensional.

Exercise 5.2 Nineteen copies of an aircraft engine part had the following running times to failures:
65 78 114 270 310 326 394 51 83 93 102 139 158 233 249 276 285 286 301.
Estimate the 20% percentile in the simplest possible way and bootstrap analyse the estimate.

Exercise 5.3 Continuation of Exercise 5.2. Data on a somewhat similar kind of unit has earlier been well fitted by the Weibull distribution $f(x) = ax^{b-1}e^{-(a/b)x^b}$. Use this information and the data above to estimate the percentile by a parametric method and bootstrap analyse the estimate. This requires a numerical solution of an equation.

Exercise 5.4 Continuation of Exercises 5.2 and 5.3. We can be more realistic by mixing times to failure with running times without failures. Let us split up the 19 values given earlier in the following way.

Times to failure: 65 78 114 270 310 326 394;

Running times without failures:
51 83 93 102 139 158 233 249 276 285 286 301.

These censored data are real and were taken from a hydrome-chanical device.

a. Find a method to estimate the 20% percentile.

b. Given some estimation method, construct a possible boot-strap analysis of the estimator.

Hint: An empirical distribution for censored data is sug-gested in the appendix of Chapter 7.

Exercise 5.5 Estimation of correlation. This exercise ex-tends the bootstrap to two-dimensional observations.

From a set of data $(x_1, y_1), \ldots, (x_n, y_n)$ the correlation is estimated by the formula

$$ r = \frac{\sum(x_i - \bar{x})(y_i - \bar{y})}{\sqrt{\sum(x_i - \bar{x})^2 \sum(y_i - \bar{y})^2}}. $$

Suppose the data can be regarded as independent, identically distributed random vectors (X, Y), with a theoretical corre-lation $\rho = E[(X - \mu_X)(Y - \mu_Y)]/(\sigma_X \sigma_Y)$. The estimator r can then be bootstrapped. Define the empirical distribu-tion (not necessarily by the distribution function), and its parameter $\tilde{\rho}$. Also tell how bootstrap values r^* are gener-ated. For a computer exercise, use for example two columns of the data in Table 5.5.

An article by Diaconis and Efron in Scientific American (1983), gives a popular illustration of bootstrapping an esti-mated correlation coefficient.

Exercise 5.6 Estimation of eigenvectors.

In Table 5.5 a group of 30 students have given their at-titudes to four questions in a course evaluation. We as-sume that students answer independently from each other (although this is by no means obvious), but answers from the

Table 5.5 *Course evaluation, attitudes graded 1 to 5.*

Person	Question 1	Question 2	Question 3	Question 4
1	5	5	5	5
2	4	4	5	5
3	4	4	5	3
4	4	4	4	3
5	4	4	3	5
6	2	3	5	1
7	3	4	4	3
8	4	3	3	2
9	3	3	3	3
10	4	4	5	3
11	4	3	4	2
12	4	4	5	3
13	4	4	4	3
14	4	4	2	3
15	5	5	3	3
16	5	5	3	2
17	5	5	4	2
18	4	4	3	4
19	5	5	4	4
20	4	4	2	3
21	5	5	3	3
22	5	5	4	1
23	5	5	4	4
24	5	5	3	4
25	4	3	4	2
26	4	4	4	3
27	5	5	5	4
28	1	2	3	1
29	3	3	3	3
30	3	4	2	5

same student to the four questions are probably dependent and produce a random vector with a 4×4 covariance matrix.

Computing the eigenvectors of covariance matrices is the core of some statistical methods like principal components analysis and factor analysis. By using some orthogonality properties of the eigenvectors, a random vector with that covariance matrix can be transformed into a vector of uncorrelated variables having certain optimal properties. Analyses of this kind are often used on data from psychological studies, aptitude tests, etc. Since the true covariances of the model are usually unknown, estimated covariance matrices are used instead. Experience shows that the results are easily stressed too hard, unless a realistic picture of the uncertainty is also given.

This exercise presumes that the reader has access to computer routines for eigenvalues and eigenvectors. Such routines exist in many standard packages such as NAG, MAT-LAB, GAUSS, and MINITAB. Use the given data. Estimate the eigenvectors of the covariance matrix, perform a bootstrap analysis, and find a way to illustrate (perhaps graphically) the uncertainties of the vectors corresponding to the two largest eigenvalues.

CHAPTER 6

Further bootstrap results

In this chapter we first collect some material necessary for the comparison of asymptotic properties and for the development of a more theoretical view of the bootstrap ideas. Then we discuss some more specific confidence interval methods for independent and identically distributed data. In the last sections we leave the identical distributions by considering regression problems, and also give up independence by considering some methodology for stochastic processes. We start off by a generalization of the bootstrap distribution.

6.1 Parametric bootstrap

The original bootstrap is non-parametric in the sense that it uses the empirical distribution $F_n(x)$. This estimate of the distribution is not based on parametric assumptions, even if parameters are usually involved in other parts of the analysis.

Suppose we know that the observations $x_1, \ldots, x_n$ are from a particular parametric class of distributions. It is then natural to change views and replace the empirical distribution by a member of the class. Efron (1982) mentions this possibility and interprets the old theory by Fisher for the standard error of a maximum likelihood estimate as a parametric bootstrap where the simulations can be replaced by analytical results. Since then the parametric version has been much investigated in research and some methods, like for instance the confidence construction suggested by DiCiccio and Tibshirani (1987), are designed for this situation only.

As substitute for the empirical distribution we typically search the distribution in the parametric class which in some sense comes closest to $F_n(x)$ or, equivalently, closest to the data. We can regard this as a projection of F_n onto the parametric class.

For the projection we need measures of closeness or distance between distributions. One such useful measure is the (log) likelihood itself. The larger it becomes, the closer we are to the data. This criterion is convenient when we intend to estimate parameters by maximum likelihood anyhow, and then bootstrap in order to study their properties.

An often used measure of closeness is Kullback–Leibler's measure, which is well known from information theory (Kullback 1968). This measure is sometimes called entropy and can be defined in the following way. Let f be a density and G a distribution function. Set

$$
\begin{aligned}
KL(f,G) &= \int \ln(f(x))dG(x) \\
&= \begin{cases} \int \ln(f(x))g(x)dx & \text{if } G \text{ continuous} \\ \sum_i \ln(f(x_i))g(x_i) & \text{if } G \text{ discrete,} \end{cases}
\end{aligned} \quad (6.1)
$$

where $g(x) = G'(x)$ in the continuous case and where G has jumps of size $g(x_i)$ at x_i in the discrete case. The last version is also useful when $f(x_i)$ represents discrete probabilities. When f and g are of the same type one can show that, for G and g fixed, the measure is maximized when $f = g$. See Kullback (1968) and Rao (1973, p. 58–59). For the empirical distribution we get

$$
KL(f, F_n) = \sum \ln f(x_i) \cdot \frac{1}{n}
$$

since F_n has jumps of size $\frac{1}{n}$ at each observation. This is the average log likelihood, or the ordinary log likelihood divided by n. This means that likelihood maximization and KL maximization are equivalent in this situation with independent data. By both methods we will resample from the distribution with maximum likelihood estimated parameters.

Alternatives to the likelihood measure can be constructed from other estimation methods and from test variables for model fit. One such possibility, the Chi-square test variable, is used in the second example below. Other distance measures are given by Kolmogorov-Smirnov's test variable and the like.

Example 6.1   Normal distribution
A sample $x_1, \ldots, x_n$ is given. Let the set of normal distributions with densities

$$f(x; \mu, \sigma) = \frac{1}{\sigma\sqrt{2\pi}} e^{-\frac{(x-\mu)^2}{2\sigma^2}} \tag{6.2}$$

be the parametric class of interest. Suppose $\theta = g(\mu, \sigma)$ is the parameter we want to estimate. One natural estimator is $\hat{\theta} = g(\bar{x}, s)$, where we simply plug in the traditional estimates of μ and σ.

We will now bootstrap from the density closest to the data. By the log likelihood, or equivalently Kullback-Leibler's measure, the density with maximum likelihood estimated parameters is the closest. Parametric resampling is therefore made from the density $f(x; \hat{\mu}, \hat{\sigma})$, with $\hat{\mu} = \bar{x}$, and $\hat{\sigma} = \sqrt{\sum(x_i - \bar{x})^2/n} = s\sqrt{1 - 1/n}$.

If $X_1^*, \ldots, X_n^*$ is a bootstrap sample, we get the bootstrap estimate $\hat{\theta}^* = g(\bar{X}^*, s^*)$. Repeated estimates are compared to the true parameter $\tilde{\theta} = g(\hat{\mu}, \hat{\sigma})$ of the bootstrap distribution. Whether we actually perform simulations, or use known properties of $\bar{X}$, s for a more theoretical computation is purely a matter of personal preference.

In our next example we demonstrate the use of a different measure of the distance between a parametric distribution and the data.

Example 6.2
Let 1 1 3^+ 1 2 1 2 2 3^+ 1 3^+ 1 3^+ 2   be data from a Poisson distribution with parameter λ. The data are censored so that only non-zero data are registered, and results ≥ 3 are grouped in 3^+. We are then led to the following parametric probabilities for 1, 2, and 3^+

$$p_1 = P(X = 1|X \geq 1) = \frac{\lambda e^{-\lambda}}{1 - e^{-\lambda}}$$

$$p_2 = P(X = 2|X \geq 1) = \frac{(\lambda^2/2)e^{-\lambda}}{1 - e^{-\lambda}} \qquad (6.3)$$

$$p_3 = P(X \geq 3|X \geq 1) = 1 - p_1 - p_2.$$

In contrast to the example above, we now choose a Chi-square test variable to measure the distance between this parametric distribution and the data. There are five, four, and three observations of type 1, 2, and 3^+. Let

$$T(\lambda) = \frac{(5 - 12p_1)^2}{12p_1} + \frac{(4 - 12p_2)^2}{12p_2} + \frac{(3 - 12p_3)^2}{12p_3}.$$

This distance measure is minimized for $\hat{\lambda} = 1.53$. Parametric resampling can now be made from the three point distribution

$$(p_1, p_2, p_3) = (0.423, 0.323, 0.254)$$

defined by (6.3) with $\lambda = \hat{\lambda}$. It can of course also be made from the Poisson distribution with parameter 1.53 followed by censoring and grouping.

   In this case the parametric bootstrap comes very close to the non-parametric version, which uses the observed frequencies $(5/12, 4/12, 3/12) = (0.417, 0.333, 0.250)$ in the resampling. With another set of data, the methods can be very different. If for example we observe 1 1 3^+ 1 3^+ 1 1 3^+ 3^+ 1 1 3^+, we will estimate $\hat{\lambda} = 1.65$. Parametric bootstrap then gives resampling with the probabilities $(0.392, 0.323, 0.285)$, but the non-parametric version uses $(0.583, 0, 0.417)$ since no observations of 2 are made.

Remarks
When all the estimation and resampling is based on maximum likelihood and/or the KL-measure, the bootstrap re-

duces to a pure simulation of the estimation problem for the distribution defined by the estimated parameters. Such simulations are not new. They have in fact been made long before the invention of bootstrap methods.

The parametric bootstrap with Kullback-Leibler's measure brings traditional likelihood theory and bootstrap methods close to each other. Some theoretical properties and asymptotic results are therefore better studied for this method. Setting up a class of parametric distributions requires of course a lot of external knowledge, which is sometimes or often not available. Many problems also have good classical solutions when a parametric likelihood can be written down. The practical need for this kind of bootstrap is therefore perhaps somewhat limited by these facts.

6.2 Basic asymptotic concepts

Asymptotic results and hints about such results have not been invisible before this section, but we will give a more valid discussion of such material in this and the following sections. We start with some classical results and some terminology.

6.2.1 Boundedness, $\sqrt{n}$-consistency, and order of convergence

We need some basic definitions of convergence for random sequences.

A sequence Z_n of stochastic variables *converges in probability* to 0 if for any $\varepsilon > 0$, $\delta > 0$ there exists n_0 such that $P(|Z_n| > \delta) < \varepsilon$ for $n > n_0$. The sequence converges in probability to a random variable Y if $Z_n - Y$ converges to 0 in probability.

A more demanding definition is the following. The sequence Z_n *converges with probability* 1 to Y if $P(|Z_n - Y| \to 0) = 1$. More explicitly this means that for every outcome ω, except in a set of probability 0, and for every $\delta > 0$ there exists $N(\omega, \delta) < \infty$ such that $|Z_n - Y| < \delta$ for $n > N(\omega, \delta)$.

A weaker definition is also possible by saying that the sequence Z_n *converges in distribution* to Y if the distribution functions $F_{Z_n}(x) \to F_Y(x)$ for every x where $F_Y(x)$ is continuous. It is convenient, but somewhat misleading, to refer to the variables in this definition. Actually it is a convergence of distribution functions without any condition on $|Z_n - Y|$ being small, and the limit variable Y can for example be independent of the sequence Z_n.

Convergence with probability 1 implies convergence in probability, and convergence in probability implies convergence in distribution. See for example Chung (1974). We will now go on to definitions of boundedness and consistency.

The sequence Z_n is *stochastically bounded* if for every $\varepsilon > 0$ there exists numbers M and n_0 such that

$$P(|Z_n| > M) < \varepsilon, \quad \text{for } n > n_0. \tag{6.4}$$

A sufficient condition for this is that Z_n converges in distribution.

If $\mathbf{Z}_n$ is vector valued with components Z_{ni}, we can define a norm $\|\mathbf{Z}_n\|$ which replaces $|Z_n|$ in the definition. The Euclidean norm $\sqrt{\mathbf{Z}_n' \mathbf{Z}_n}$ or the maximum norm $\max_i |Z_{ni}|$ both have the property that $\mathbf{Z}_n$ is stochastically bounded if and only if every component is so. The same kind of extension to vector variables applies to convergence in probability and convergence with probability 1.

A sequence of estimators $\hat{\eta}_n$ of the parameter η is *weakly consistent* if $\hat{\eta}_n \to \eta$ in probability and *strongly consistent* if the convergence is with probability 1.

The sequence $\hat{\eta}_n$ is $\sqrt{n}$-*consistent* if the sequence $\mathbf{Z}_n = \sqrt{n}(\hat{\eta}_n - \eta)$ is stochastically bounded. This is a typical behaviour for most of the traditional estimates.

A more detailed description of asymptotic behaviour is also possible since many estimates and other functions of the data have distribution functions which can be expanded as series in powers of $1/\sqrt{n}$, where n is the sample size. See for instance the presentation of Edgeworth series later in this chapter. Let

$$G_n(x;\eta) = G_0(x;\eta) + \frac{1}{\sqrt{n}}g_{n1}(x;\eta) + \frac{1}{n}g_{n2}(x;\eta) + \ldots$$

be such an expansion of the distribution where η is the parameter of the problem. If the estimate, say, is defined so that $G_n(x;\eta)$ converges to $\Phi(x)$, the standard normal, or some other η-independent $G_0(x)$, we have

$$G_n(x;\eta) = G_0(x) + \frac{1}{\sqrt{n}}g_{n1}(x;\eta) + \frac{1}{n}g_{n2}(x;\eta) + \ldots \quad (6.5)$$

By ingenious definitions of the functions some of the lower order terms can be removed from the expansions, and higher order convergence to $G_0(x)$ is then achieved. This is the goal of much asymptotic work for bootstrapped, jackknifed (Miller 1974), and pivoted statistics. See for example Singh (1981), Babu and Singh (1983, 1985), Hall (1986, 1988), Efron (1987), and in particular Hall (1992). The lowest power of $1/\sqrt{n}$ remaining in (6.5) is called the *order of the convergence*. Instead of the distribution function we may study the endpoints of confidence intervals, or compare other probabilities or estimates. If the errors can be expanded in $1/\sqrt{n}$-series we will use the corresponding definition of convergence order.

Of course, asymptotic results will not tell the whole story. Removing lower order terms can lead to higher variances for small samples, and larger coefficients for the terms remaining in the expansion. Simulations or special analyses are therefore needed in order to check small sample properties. Usually such studies, together with the computational efforts, set a practical limit so that no more than one or two powers of $1/\sqrt{n}$ should be removed.

6.2.2   Ordo terms

We need a comfortable description of the errors when we
have an improving sequence of approximations. Usually the
relative speed with which the approximations improve is relatively easy to find, but a useful bound on the absolute errors
can be too hard to find or too dependent on the fine details
of each problem. Ordo terms are convenient for describing
the speed of convergence.

The deterministic ordo definitions are given by $a_n = o(b_n)$
if $a_n/b_n \to 0$ and $c_n = O(b_n)$ if c_n/b_n is bounded as $n \to \infty$.
These definitions have the following random analogues.

$X_n = o_p(b_n)$ if $X_n/b_n \to 0$ in probability, and $Y_n = O_p(b_n)$
if Y_n/b_n is stochastically bounded.

The usual rules $o(b_n) + o(b_n) = o(b_n)$, and $o(a_n) + o(b_n) =
o(b_n)$ if $\lim a_n/b_n < k < \infty$ allow us to collect several small
terms in a single one. Also $o(b_n)$ is $o_p(b_n)$. All these rules
are equally valid for big ordo.

6.2.3   A reminder of some classical limit theorems

For reference, some basic theorems are listed here. We give
no proofs, since they are easily found in standard texts in
probability like Feller (1971), Chung (1974), and the interested reader can look them up there.

First two versions of the central limit theorem, one multivariate which of course covers the univariate case as well,
then a more precise result in the univariate case.

Theorem 6.1   A multivariate central limit theorem
Let $\mathbf{X}_1, \mathbf{X}_2, \ldots$ be independent, identically distributed m-
dimensional vectors with $E[\mathbf{X}_i] = \mathbf{0}$ and finite covariance
matrix $E[\mathbf{X}_i\mathbf{X}_i'] = \mathbf{C}$. Then, as $n \to \infty$, $(\mathbf{X}_1 + \ldots +
\mathbf{X}_n)/\sqrt{n}$ has asymptotically an m-dimensional normal dis-
tribution with mean zero and covariance matrix $\mathbf{C}$, in short
$N_m(\mathbf{0}, \mathbf{C})$.

See for example Feller (1971, p 260).

The formulation we give of the central limit theorem below takes an extra condition in terms of third moments and gives in return a uniform bound for finite samples from all such distributions. The theorem was found (reported) during the Second World War independently by Berry in 1941 and Esséen in 1942.

Theorem 6.2   The Berry-Esséen central limit theorem
If $X_1, X_2, \ldots$ are independent, identically distributed scalar variables with $E[X_i] = 0$, $E[X_i^2] = \sigma^2$, and $\rho = E[|X_i|^3] < \infty$, then for all x and n

$$\left| P(\frac{X_1 + \ldots + X_n}{\sigma\sqrt{n}} \leq x) - \Phi(x) \right| < K \frac{\rho}{\sigma^3 \sqrt{n}}, \qquad (6.6)$$

where K is a universal constant.

The proof with $K = 3$ can be found in Feller (1971, p 542–544).

Theorem 6.3   Borel-Cantelli's lemma
Let A_i be a denumerable set of events in the sample space Ω. Let $N = N(\omega) = \#\{A_i; \omega \in A_i\} \leq \infty$ count the number of events that occur for the outcome ω. If $\sum P(A_i) < \infty$ then $P(N = \infty) = 0$.

Intuitively, let B be the set where infinitely many A_i occur. If $P(B) > 0$, then this probability must be counted infinitely often in $\sum P(A_i)$ and the sum diverges. A strict proof is given in Chung (1974, p 73). A reverse of the theorem exists in the special case of independent events A_i.

Theorem 6.4   Strong law of large numbers
If $X_1, X_2, \ldots$ are independent variables with a common distribution $F(x)$ and finite mean $\mu = E[X_i] = \int x \, dF(x)$ then $\overline{X}_n = \frac{1}{n}(X_1 + \ldots + X_n) \to \mu$ with probability 1 as $n \to \infty$.

Corollary

If $E[X_i^k] = \mu_k$ is finite then $\frac{1}{n}\Sigma_1^n X_i^k \to \mu_k$ with probability 1 and $\frac{1}{n}\Sigma|X_i|^k \to E[|X_i|^k]$ also with probability 1. The last expectation is finite when μ_k is finite.

The strong law (and more general versions of it) is proved in Feller (1971, p. 238–240). The corollary is an obvious consequence but can be useful for reference.

6.3 Convergence of resampling distributions

The bootstrap sample is drawn from the empirical distribution $F_n(x)$ (or some parametric or other version of this concept) and the true sample comes from the distribution $F(x)$. The first requirement for correct asymptotic results is therefore that $F_n(x)$ converges to $F(x)$. Fortunately such a result is generally available for independent and identically distributed data. However, this is not enough. We also need convergence at sufficient speed for asymptotically correct confidence intervals and for useful approximations for reasonable finite sample sizes. The convergence of F_n must also carry over to the parameter being studied. Therefore, the parameter and the estimate have to be sufficiently smooth functions of the distribution and the data so that the differences, which are always there, do not get dramatically magnified.

6.3.1 The finite case

The first result of this kind was given by Efron in his pioneering paper. See Efron (1979, 1982). It was given for the finite case only, but the result has some relevance also for non-finite situations when a finite case can provide a close approximation. Efron proved that if a variable can take values in a finite set $\{a_1, \ldots, a_m\}$ only, and we have independent and identically distributed data on such variables, then the bootstrap has asymptotically the correct distribution for a large class of problems.

The idea of the proof is quite simple. Let $\mathbf{f} = (f_1, \ldots, f_m)'$ be the vector of probabilities for the possible values.

Take n random observations $X_1, X_2, \ldots, X_n$ and let $\hat{\mathbf{f}} = \frac{1}{n}(\mathbf{Z}_1 + \ldots + \mathbf{Z}_n)$ be the observed relative frequencies of the values $\{a_1, \ldots, a_m\}$. Each $\mathbf{Z}_i$ is a vector with 1 at the position of the data X_i and 0 at the other positions. Since $E[\mathbf{Z}_i] = \mathbf{f}$, the strong law of large numbers (Theorem 6.4) gives that $\hat{\mathbf{f}} \to \mathbf{f}$ with probability 1 when $n \to \infty$.

For a given n and $\hat{\mathbf{f}}$, let $\hat{\mathbf{f}}^*$ be the relative frequencies from n resampled data drawn with the probability vector $\hat{\mathbf{f}}$. We have the results

$$E[\hat{\mathbf{f}} - \mathbf{f}] = 0, \qquad E[\hat{\mathbf{f}}^* - \hat{\mathbf{f}}|\hat{\mathbf{f}}] = 0,$$

where the second expectation expresses the same thing as the first, but for the bootstrap resample. Also the unconditional result $E[\hat{\mathbf{f}}^* - \hat{\mathbf{f}}] = 0$ follows from the same conditional result.

For the second moments we can compute

$$E[(\hat{\mathbf{f}} - \mathbf{f})(\hat{\mathbf{f}} - \mathbf{f})'] = \frac{1}{n}\mathbf{C},$$

where the diagonal matrix element $c_{ii} = f_i(1 - f_i)$ is the variance of a zero-one variable, and $c_{ij} = -f_i f_j$, $i \neq j$. In the same way

$$E[(\hat{\mathbf{f}}^* - \hat{\mathbf{f}})(\hat{\mathbf{f}}^* - \hat{\mathbf{f}})'|\hat{\mathbf{f}}] = \frac{1}{n}\hat{\mathbf{C}},$$

where $\hat{\mathbf{C}}$ has elements like $\mathbf{C}$ but with $\hat{\mathbf{f}}$ replacing $\mathbf{f}$. By the multivariate version of the central limit theorem, (Theorem 6.1), we have the following convergence in distribution

$$\sqrt{n}(\hat{\mathbf{f}} - \mathbf{f}) \to N_m(0, \mathbf{C}),$$

and given $\hat{\mathbf{f}}$,

$$\sqrt{n}(\hat{\mathbf{f}}^* - \hat{\mathbf{f}}) \to N_m(0, \hat{\mathbf{C}}). \tag{6.7}$$

(The last convergence is uniform for $\hat{\mathbf{f}}$ in a neighbourhood of the true $\mathbf{f}$).

Since $\hat{\mathbf{f}} \to \mathbf{f}$ with probability 1, it follows that $\hat{\mathbf{C}} \to \mathbf{C}$ with the same probability, and the resample will asymptotically show up the distribution of the original data with the corresponding random fluctuations around (to the order of $1/\sqrt{n}$).

But the observations were $X_1, \ldots, X_n$ and not $\hat{\mathbf{f}}$. The analysis therefore covers such functions of $(X_1, \ldots, X_n)$ which can be interpreted as functions of $\hat{\mathbf{f}}$ and consequently can be resampled by the same functions of $\hat{\mathbf{f}}^*$. Now, if a function $g(X_1, \ldots, X_n)$ is invariant under permutations of the X_i, we can choose a permutation such that all $X_i = a_1$ come first, followed by all $X_i = a_2$ and so on. Let $X_{k_1}, \ldots, X_{k_n}$ be this permuted vector. Then $f_1, f_2, \ldots, f_m$ gives full information about this permuted data vector. Consequently, for the permutation invariant case we have $g(X_1, \ldots, X_n) = g(X_{k_1}, \ldots, X_{k_n}) = h(\hat{\mathbf{f}})$ for some function h. Conversely it is easily seen that any function of $\hat{\mathbf{f}}$ corresponds to a permutation invariant function of the data.

Most quantities of interest in the independent and identically distributed case are in fact invariant under permutations of the data. Consequently, they are functions of $\hat{\mathbf{f}}$. One last condition defines the useful class of problems. Provided the studied parameter θ is a sufficiently smooth function of $\mathbf{f}$ and the function $h(\hat{\mathbf{f}}^*)$ we study in the resampling is also smooth, the similarities between $\hat{\mathbf{f}}^* - \hat{\mathbf{f}}$ and $\hat{\mathbf{f}} - \mathbf{f}$ will carry over to these functions. (Functions with bounded derivatives in the region of interest will do.) For such smooth and permutation invariant functions and variables with finite sets of values the asymptotic validity of the bootstrap method is now proved.

6.3.2 A continuous continuation

The finite case treated by Efron soon got followers for various non-finite situations. The natural extension to the asymptotic normal distribution of the relative frequencies then became a stochastic process with normal distributions. More precisely a Brownian bridge process arose. This process can be defined from the more basic Wiener process. We therefore introduce these processes first.

A *Wiener process* is a continuous normal (Gaussian) process $W(t), 0 \leq t$, such that $W(0) = 0$, $E[W(t)] = 0$, and $E[W(t)W(u)] = \sigma^2 \min(t, u)$ for $t, u \geq 0$. This characterizes all distributions of the process, since normal distributions are determined by their means and covariances. A characteristic property is that the Wiener process has independent increments such that $W(t) - W(u)$ is $N(0, \sigma\sqrt{|t - u|})$. We will standardize on $\sigma = 1$ here, so in particular $W(t)$ becomes $N(0, \sqrt{t})$. The independent normal increments make the Wiener process an excellent model for continuous random walks and diffusion.

A *Brownian bridge* is another Gaussian process $B(t)$, $0 \leq t \leq 1$, which can be represented as

$$B(t) = W(t) - tW(1), \quad 0 \leq t \leq 1. \tag{6.8}$$

We have here defined the process on the unit time interval and with the standardized Wiener process ($\sigma = 1$). Other definitions are possible but can be seen as simple rescalings of this process. It is easily seen that $B(0) = B(1) = 0$, $E[B(t)] = 0$, and $E[B(t)B(u)] = \min(t, u) - tu$. For $0 < t < u < 1$ the last expression becomes $t(1 - u)$, the product of the distances to the two end points of the time interval.

We will now turn our attention to the empirical distributions and the bootstrap analysis. Let $F_n(t)$ be the empirical distribution of n independent variables $X_1, \ldots, X_n$ from the distribution $F(x)$. Also let $F_n^*(t)$ be the corresponding empirical distribution of the bootstrap resample $X_1^*, \ldots, X_n^*$. Both $F_n(t)$ and $F_n^*(t)$ are here considered as random objects (processes) parametrized by t, $-\infty < t < \infty$. From Bickel and Freedman (1981) we have the following result asymptotically in n.

Theorem 6.5

For almost every sequence $X_1, X_2, \ldots$, and conditional on $X_1, \ldots, X_n$, the process

$$Z_n(t) = \sqrt{n}(F_n^*(t) - F_n(t)) \tag{6.9}$$

converges weakly to $B(F(t))$, the Brownian bridge with the distribution function as parameter.

The result is proved without any extra conditions on the distribution $F(t)$. (Almost every sequence means that with probability 1 we will have such a sequence, and we can hardly ask for more). The exact nature of the weak convergence is described in Bickel and Freedman's work, and more completely in Billingsley (1968).

The same, but unconditional, convergence to $B(F(t))$ is valid also for $\sqrt{n}(F_n(t) - F(t))$, but this is an older and well-known result.

Just like Efron's result for the finite case, this result gives a feel for the asymptotic validity of the bootstrap procedure in many situations where the studied parameters and functions of the data are smooth functions of the true and empirical distributions. But there are also more direct uses. Following Bickel and Freedman once more we may set confidence bounds on the distribution function in the following way.

Let $0 < \alpha < 1$ and choose a value $c = c(F_n)$ from the (simulated) bootstrap distribution so that

$$P_*(\sqrt{n} \sup_x |F_n^*(x) - F_n(x)| \le c) \to 1 - \alpha.$$

Then, since $\sqrt{n}(F_n^*(x) - F_n(x))$ and $\sqrt{n}(F_n(x) - F(x))$ converge to the same process $B(F(t))$, we have that also

$$P(\sqrt{n} \sup_x |F_n(x) - F(x)| \le c) \to 1 - \alpha.$$

An asymptotically correct confidence band for $F(t)$ is therefore given as $F(x) = F_n(x) \pm c/\sqrt{n}, \ -\infty < x < \infty$.

For distributions with jumps, this interval can improve on the more analytical Kolmogorov-Smirnov solution of the same problem. However, the bootstrap solution can also easily extend to other versions of the problem. We can for example bootstrap an expression like

$$\sup_x \frac{\sqrt{n}|F_n^*(x) - F_n(x)|}{a + \sqrt{F_n^*(x)(1 - F_n^*(x))}}$$

where $a > 0$ is a protection against division by zero and against blowing up the differences. A $1 - \alpha$ upper bound c for this bootstrap distribution leads to the confidence band $F(x) = F_n(x) \pm c(a + \sqrt{F_n(x)(1 - F_n(x))})/\sqrt{n}$ with varying width. The convergence above together with $F_n^* \to F_n$ and $F_n \to F$ gives the asymptotic validity.

Further advanced results in this direction have been developed by several authors. We refer the interested reader to Csörgö and Mason (1989) and references therein.

6.4 Asymptotic results for averages and percentiles

In this section we will describe some more exact results stated by Singh (1981). We have modified the presentation somewhat in order to give some proofs without too extensive background. Let $X_1, X_2, \ldots$ be a sequence of independent variables with common distribution $F(x)$, expected values μ and standard deviations σ. Use X as a general symbol of a variable with the same distribution F. Take the first n variables from the sequence and imagine a bootstrap resample $X_1^*, \ldots, X_n^*$ from the outcome of the original variables. Introduce the notation

$$\overline{X}_n = \frac{X_1 + \ldots + X_n}{n} \qquad \tilde{\mu} = \overline{X}_n \qquad \tilde{\sigma} = \sqrt{\frac{\sum_1^n (X_i - \overline{X}_n)^2}{n}}$$

$$\overline{X}_n^* = \frac{X_1^* + \ldots + X_n^*}{n}$$

$$\rho = E[|X - \mu|^3] \qquad \tilde{\rho} = E_*[|X^* - \tilde{\mu}|^3] = \frac{\sum |X_i - \overline{X}_n|^3}{n}.$$

All the symbols $\tilde{\mu}$, $\tilde{\sigma}$, $\tilde{\rho}$ are functions of the original sample and of course dependent on the sample size n, but we suppress this in our notation.

6.4.1 Averages

The nice asymptotic behaviour of sample averages is given by the following theorem, which is then sharpened in the next two theorems.

Theorem 6.6
If $\rho < \infty$ then with probability 1 the sequence $X_1, X_2, \ldots$ is such that

$$\sup_x |P(\sqrt{n}(\overline{X}_n - \mu) \le x) - P_*(\sqrt{n}(\overline{X}_n^* - \tilde{\mu}) \le x)| \to 0 \quad (6.10)$$

when $n \to \infty$.

Proof By the Berry-Esséen central limit theorem, Theorem 6.2, we have

$$|P(\sqrt{n}(\overline{X}_n - \mu) \le x) - \Phi(\frac{x}{\sigma})| < K \frac{\rho}{\sigma^3 \sqrt{n}} \quad (6.11)$$

$$|P_*(\sqrt{n}(\overline{X}_n^* - \tilde{\mu}) \le x) - \Phi(\frac{x}{\tilde{\sigma}})| < K \frac{\tilde{\rho}}{\tilde{\sigma}^3 \sqrt{n}}. \quad (6.12)$$

Let $n \to \infty$. By the strong law of large numbers, Theorem 6.4 with Corollary, we have $\tilde{\mu} \to \mu$ with probability 1 if $E[X]$ is finite, $\tilde{\sigma} \to \sigma$ if $E[X^2]$ (or σ^2) is finite, and $\tilde{\rho} \to \rho$ if $E[|X|^3]$ (or $E[X^3]$ or ρ) is finite, all with probability 1. Also ρ finite implies finite μ and σ. Since the compared distributions in (6.10) come close to the same normal distribution, the result follows.

Comment
Singh (1981) and Bickel and Friedman (1981) both prove
the result using finite second moments only. That the left
member of (6.11) converges to zero is in fact a standard result
in central limit theory, and we will take it as given. The
result will therefore follow if we can show the next theorem
and apply the result to the inequality (6.12).

Theorem 6.7   Relaxation of third moment condition
$E[X^2] < \infty$ implies that $\tilde{\rho}/\sqrt{n} \to 0$.

Proof First we have $\tilde{\rho} = E_*[|X^* - \tilde{\mu}|^3]$, and the algebraic
inequality $|a - b|^3 \le (|a| + |b|)^3 \le 4|a|^3 + 4|b|^3$ implies that

$$\tilde{\rho} \le 4(E_*[|X^*|^3] + |\tilde{\mu}|^3) = 4(\frac{1}{n}\sum_1^n |X_i|^3 + |\overline{X}_n|^3).$$

Divide this by $\sqrt{n}$. Since $\overline{X}_n \to \mu$ with probability 1,
$4|\overline{X}_n|^3/\sqrt{n} \to 0$ and causes no difficulty. The interesting
part of $\tilde{\rho}/\sqrt{n}$ is therefore $\frac{1}{n\sqrt{n}}\Sigma_1^n|X_i|^3$. Now choose an $\varepsilon > 0$
and regard the sum

$$\sum_{i=1}^{\infty} P(|X_i| \ge \sqrt{i}\varepsilon) = \sum_{i=1}^{\infty} P(X_i^2 \ge i\varepsilon^2)$$
$$= \sum_{i=1}^{\infty} iP(i\varepsilon^2 \le X_i^2 < (i+1)\varepsilon^2) \le \frac{E[X_i^2] + \varepsilon^2}{\varepsilon^2}, \quad (6.13)$$

which is finite if $E[X^2] < \infty$.

Here the second equality follows by checking how many
times each subinterval of the third member is contained in
the terms of the second sum, and the next approximation
with an expected value is a standard but useful one.

By the Borel–Cantelli lemma, Theorem 6.3, we find that
with probability 1 only finitely many of the events $\{|X_i| \ge \sqrt{i}\varepsilon\}$ will occur, and there will thus be a last such event for

each such case. Take a typical (probability 1) outcome ω such that $\frac{1}{n}\Sigma_1^n X_i^2 \to E[X^2]$ and $N = N(\omega) = \sup\{i; \frac{|X_i|}{\sqrt{i}} \geq \varepsilon\}$ is finite. For $n > N$ we have

$$\frac{\sum_{i=1}^{n} |X_i|^3}{n\sqrt{n}} = \frac{\sum_{i=1}^{N} |X_i|^3}{n\sqrt{n}} + \frac{1}{n} \sum_{N+1}^{n} \frac{|X_i|^3}{\sqrt{n}}$$

$$\leq \frac{\sum_{i=1}^{N} |X_i|^3}{n\sqrt{n}} + \frac{1}{n} \sum_{N+1}^{n} |X_i|^2 \varepsilon.$$

As $n \to \infty$, N and X_i stay fixed for each given ω. The first term of the upper bound therefore converges to zero, and the second to $\varepsilon E[X^2]$. But ε is arbitrarily small, so the convergence must be to zero with probability 1 and the proof is finished.

The next result shows that we may standardize by dividing by the standard deviation without destroying the comparison. In fact this is typically beneficial as we have indicated elsewhere. We will also get information about the speed of convergence.

Theorem 6.8 (Singh)
If X_i are independent and identically distributed with $\rho < \infty$ then

$$\limsup_{n\to\infty} \frac{\sqrt{n}\sigma^3}{\rho} \sup_x \left| P(\frac{\overline{X}_n - \mu}{\sigma/\sqrt{n}} \leq x) - P^*(\frac{\overline{X}_n^* - \tilde{\mu}_n}{\tilde{\sigma}_n/\sqrt{n}} \leq x) \right|$$
$$\leq 2K \qquad (6.14)$$

where K is the Berry–Esséen constant.

Proof As in the preceding proof we use the Berry–Esséen theorem to compare both distributions with $\Phi(x)$ and the result follows directly from the convergence of $\tilde{\sigma}$ and $\tilde{\rho}$ to σ and ρ.

With more effort Singh shows that the limit in (6.14) is actually zero if X_i have a continuous distribution (non-lattice is enough), and becomes $h/\sqrt{2\pi\sigma^2}$ if X_i take values in the set $0, \pm h, \pm 2h, \ldots$ only.

Without division by the standard deviation he also obtains as comparison that for $E[X^4] < \infty$

$$\limsup_{n\to\infty} \frac{\sqrt{n}}{\log\log n} \sup_x \left| P(\sqrt{n}(\overline{X}_n - \mu) \le x) \right.$$
$$\left. - P^*(\sqrt{n}(\overline{X}_n^* - \tilde\mu_n) \le x) \right| = \frac{\sqrt{E[(X_i - \mu)^4] - \sigma^4}}{2\sigma^2\sqrt{\pi e}},$$

which demonstrates a somewhat slower convergence than (6.14) for the mean without standardization by the standard deviation.

6.4.2 Median and percentile estimation

Closely related to estimating distributions is the problem of estimating percentiles. We have illustrated some elementary bootstrapping of the median in Chapter 5. The main difference in theory comes from the possibility that the distribution function may be very flat around the required percentile, which may cause both non-uniqueness and estimation problems. By proper conditions this case can be avoided and useful results stated for all other situations. The kind of results that may be obtained for percentiles is illustrated by the following theorem for medians from Bickel and Freedman (1981).

Let $X_1, X_2, \ldots$ be independent with the distribution $F(x)$ and having a unique median θ. Let $\hat\theta_n$ be the observed median of $X_1, \ldots, X_n$, with a proper definition if n is even, and let $\hat\theta_n^*$ be the observed median of the bootstrap resample $X_1^*, \ldots, X_n^*$ drawn from $F_n(x)$ in the usual way.

Theorem 6.9
If $F(x)$ has positive and continuous derivative $f(x)$ in a neighbourhood of θ, then for almost all sample sequences $X_1, X_2, \ldots$ and conditional on $X_1, \ldots, X_n$

$$\sqrt{n}(\hat{\theta}_n^* - \hat{\theta}_n) \text{ converges weakly to } N\left(0, \frac{1}{2f(\theta)}\right), \quad (6.15)$$

which is the limit (unconditional) distribution of $\sqrt{n}(\hat{\theta}_n - \theta)$. Remember that we use $N(\mu, \sigma)$ as our notation, not $N(\mu, \sigma^2)$.

We omit the proof. The theorem gives an asymptotic justification for approximating a confidence interval for θ by the observed bootstrap distribution. It also justifies some other versions where for example only the variance is estimated from the bootstrap, and the normal distribution is used, or where we estimate $f(\hat{\theta}_n)$ in some smooth consistent way and then bootstrap a pivot variable.

More exact information about the convergence of percentile estimates is given by the following theorem from Singh (1981). Let $0 < t < 1$, and let θ be the t-percentile of the distribution $F(x)$ (unique by the condition below). Then $F(\theta) = t$ or $\theta = F^{-1}(t)$. For the discrete empirical distribution we define $F_n^{-1}(t) = \sup\{x; F_n(x) \le t\}$ and take this value as our estimate $\hat{\theta}$. In the same way we define $\hat{\theta}^* = F_n^{*-1}(t)$ from the discrete empirical distribution of the bootstrap sample $X_1^*, \ldots, X_n^*$.

Theorem 6.10
If $F(x)$ has a bounded second derivative in a neighbourhood of θ and $F'(\theta) > 0$, then with probability 1

$$\limsup_{n \to \infty} \frac{n^{\frac{1}{4}} \sup_x |P(\sqrt{n}(\hat{\theta} - \theta) \le x) - P^*(\sqrt{n}(\hat{\theta}^* - \hat{\theta}) \le x)|}{\sqrt{\log \log n}}$$
$$= K_{t,F},$$

a constant depending on t and F only.

Thus the convergence is with the speed $\sqrt{\log\log n}/\sqrt{\sqrt{n}}$, a slow convergence but equivalent to that of a normal approximation which is sometimes used instead, so in a way it is typical for the problem.

6.5 Edgeworth expansion

In the central limit theorem, the distribution of a properly normalized sum of random variables converges to the normal distribution. Edgeworth expansion produces correction terms to the normal distribution for finite sums. The important feature of these corrections is that they are expanded as a series in powers of $1/\sqrt{n}$. This kind of expansion can be generalized to smooth, sufficiently differentiable functions of sums of random vectors, as shown by Bhattacharya and Ghosh (1978), and can also be inverted to a series expansion of the percentiles or confidence limits for a given probability. This generalization was exploited for the bootstrap by Hall (1988).

The central limit theorem is usually proved by some version of basically the same transformation method. Let X be random with the distribution $F(x)$, mean μ, and variance σ^2. Define the transform

$$M(t; X) = E[e^{tX}] = \int e^{tx}dF(x) = \int e^{tx}f(x)\,dx \quad (6.16)$$

where the last member is valid if $f(x) = F'(x)$ exists. With $t = i\omega$ imaginary, we get the characteristic function which is another name for the Fourier transform of the distribution. If t is always real, the $M(t; X)$ may not exist for $t \neq 0$, but if convergent, we call it the *moment generating function*. Finally, if $X \geq 0$ and we set $t = -s$, we have the Laplace transform. Any such transform, which exists in an interval, has a 1:1 relation to the distribution F.

We need to recognize our limit distribution, and when X is $N(0,1)$ we have

$$M(t; X) = \int_{-\infty}^{\infty} e^{tx} \frac{1}{\sqrt{2\pi}} e^{-x^2/2} dx = e^{\frac{t^2}{2}}. \qquad (6.17)$$

Suppose the distribution of X is at least exponentially decreasing in the tails, $P(|X| > x) < A \exp(-Bx)$ for some A and B. Then X has moments of all orders, and $M(t; X)$ exists for $-B < t < B$. Furthermore, the following interchange of expectation and differentiation is legal:

$$\frac{d^r}{dt^r} M(t; X) = \frac{d^r}{dt^r} E[e^{tX}] = E[\frac{d^r}{dt^r} e^{tX}] = E[X^r e^{tX}].$$

For $t = 0$ this generates the rth moment $E[X^r]$, which explains the name of the transform. Changing from X to the variable $X - \mu$, we now have mean zero and get centralized moments of X:

$$\frac{d^r M}{dt^r}(0; X - \mu) = \mu_r = E[(X - \mu)^r],$$

where $\mu_0 = 1$, $\mu_1 = 0$, $\mu_2 = \sigma^2$. A series expansion round $t = 0$ gives

$$M(t; X - \mu) = \sum_r \frac{t^r}{r!} \frac{d^r M}{dt^r}(0; X - \mu) = 1 + \frac{t^2}{2}\sigma^2 + \frac{t^3}{6}\mu_3 + \dots.$$

However, we will need instead an expansion of the logarithm $L(t) = \ln M(t; X - \mu)$. Now, in short notation and taking all t-derivatives of $M = M(t; X - \mu)$ at $t = 0$, we have (since $M' = 0$)

$$L(0) = \ln 1 = 0, \quad L'(0) = \frac{M'}{M} = \mu_1 = 0,$$

$$L''(0) = \frac{M''}{M} - \frac{M'^2}{M^2} = \mu_2 = \sigma^2,$$

$$L'''(0) = \frac{M'''}{M} - \frac{3M'M''}{M^2} + \frac{2M'^3}{M^3} = \mu_3$$

$$L''''(0) = \frac{M''''}{M} - \frac{4M'''M'}{M^2} - \frac{3M''^2}{M^2} + \frac{12M''M'^2}{M^3} - \frac{6M'^4}{M^4}$$

$$= \mu_4 - 3\sigma^4.$$

This gives the expansion

$$L(t) = \ln M(t; X - \mu) = \frac{t^2\sigma^2}{2} + \frac{t^3\mu_3}{6} + \frac{t^4(\mu_4 - 3\sigma^4)}{24} + O(t^5).$$

Now, let $X_1, X_2, \ldots$ be independent with the same distribution as X and define

$$Y_n = \frac{X_1 + \ldots + X_n - n\mu}{\sigma\sqrt{n}}.$$

By the central limit theorem, the moment generating function $M(t; Y_n)$ should converge to the transform $\exp(t^2/2)$ of the $N(0,1)$-distribution. Consider

$$M(t; Y_n) = E[e^{t\Sigma(X_i - \mu)/\sigma\sqrt{n}}] = E[\prod_{i=1}^{n} e^{\frac{t}{\sigma\sqrt{n}}(X_i - \mu)}].$$

Since X_i are independent, we get a product of expected values

$$M(t; Y_n) = \left(M(\frac{t}{\sigma\sqrt{n}}; X - \mu)\right)^n$$

with the logarithm

$$\ln M(t; Y_n) = nL\left(\frac{t}{\sigma\sqrt{n}}\right)$$

$$= \frac{t^2}{2} + \frac{t^3\mu_3}{6\sigma^3\sqrt{n}} + \frac{t^4}{24n}\left(\frac{\mu_4}{\sigma^4} - 3\right) + O\left(\frac{1}{n\sqrt{n}}\right).$$

From now on, error terms are in powers of $1/\sqrt{n}$ instead of t. We can somewhat arbitrarily limit t to a subinterval of $|t| < 1$ containing $t = 0$, to make this clear. The expanded logarithm gives

$$M(t; Y_n) = e^{\frac{t^2}{2}}\exp\left(\frac{1}{\sqrt{n}}\frac{t^3\mu_3}{6\sigma^3} + \frac{1}{24n}t^4\left(\frac{\mu_4}{\sigma^4} - 3\right) + O\left(\frac{1}{n\sqrt{n}}\right)\right)$$

$$= e^{\frac{t^2}{2}}\left[1 + \frac{1}{\sqrt{n}}\frac{t^3\mu_3}{6\sigma^3} + \frac{1}{n}\left(\frac{t^4}{24}\left(\frac{\mu_4}{\sigma^4} - 3\right) + \frac{1}{2}\frac{t^6\mu_3^2}{36\sigma^6}\right)\right] + O\left(\frac{1}{n\sqrt{n}}\right).$$

$$(6.18)$$

We need to transform this back to functions of x, and there is a theory about Hermitian polynomials ready for this. Defining $H_0(x) = 1$, $H_1(x) = x$, and the recursion

$$H_r(x) = xH_{r-1}(x) - (r-1)H_{r-2}(x), \qquad (6.19)$$

we get $H_2(x) = x^2 - 1$, $H_3(x) = x^3 - 3x$, $H_4(x) = x^4 - 6x^2 + 3, \ldots$.

For the first polynomials we observe that $H_1'(x) = H_0(x)$, $H_2'(x) = 2H_1(x)$, $H_3'(x) = 3H_2(x)$. Is this a general pattern? Suppose

$$H_k'(x) = kH_{k-1}(x), \qquad (6.20)$$

for $1 \le k \le r - 1$. Differentiating (6.19) we have $H'_r = H_{r-1} + xH'_{r-1} - (r-1)H'_{r-2} = H_{r-1} + x(r-1)H_{r-2} - (r-1)(r-2)H_{r-3} = rH_{r-1}$ since $xH_{r-2} - (r-2)H_{r-3}$ is H_{r-1} according to the recursion (6.19). The extension to $1 \le k \le r$ is proved, and (6.20) follows by induction for all k.

Having obtained this, we can derive the following important result,

$$\int_{-\infty}^{\infty} e^{tx} H_r(x)\phi(x)dx = e^{t^2/2}t^r. \qquad (6.21)$$

The point is that by this integral, $e^{t^2/2}t^r$ can be interpreted as the 'moment generating' transform of $H_r(x)\phi(x)$, and this is the kind of expression we have in our series expansions.

We can prove (6.21) as follows. Use the fact that the normal density $\phi(x) = \exp(-x^2/2)/\sqrt{2\pi}$ has $\phi'(x) = -x\phi(x)$.

$$\int_{-\infty}^{\infty} H_r(x)e^{tx}\frac{e^{-x^2/2}}{\sqrt{2\pi}}dx = e^{t^2/2}\int_{-\infty}^{\infty} H_r(x)\frac{e^{-(x-t)^2/2}}{\sqrt{2\pi}}dx$$

$$= e^{t^2/2}\int_{-\infty}^{\infty} H_r(t+y)\phi(y)dy = e^{t^2/2}I_r,$$

by the substitution $x - t = y$.

The integrals I_0, I_1, I_2 become $1, t, t^2$ as we can easily check for the first polynomials. Suppose $I_r = t^r$ for $0 \le r \le n - 1$. Try (6.19) on I_n to get

$$I_n = \int H_n(t+y)\phi(y)dy$$

$$= \int ((t+y)H_{n-1}(t+y) - (n-1)H_{n-2}(t+y))\phi(y)dy$$

$$= tI_{n-1} + \int y\phi(y)H_{n-1}(t+y)dy$$

$$- \int (n-1)H_{n-2}(t+y)\phi(y)dy$$

$$= t^n - \left[\phi(y)H_{n-1}(t+y)\right]_{-\infty}^{\infty} + \int \phi(y)H_{n-1}'(t+y)dy$$

$$- \int (n-1)H_{n-2}(t+y)\phi(y)dy = t^n,$$

where $\phi(y)$ cancels the polynomial at infinity for the expression integrated out by parts, and $H_{n-1}' = (n-1)H_{n-2}$ cancels the last integral. Now $I_r = t^r$, $0 \le r \le n$, and by induction this and (6.21) follows for all r.

We can now identify $M(t; Y_n)$ in (6.18) as the transform of

$$\phi(x) + \frac{\phi(x)}{\sqrt{n}} H_3(x)\frac{\mu_3}{6\sigma^3}$$

$$+ \frac{\phi(x)}{n}\left(\frac{H_4(x)}{24}(\frac{\mu_4}{\sigma^4} - 3) + H_6(x)\frac{\mu_3^2}{72\sigma^6}\right) + O(\frac{1}{n\sqrt{n}}).$$

$$(6.22)$$

This is a two term Edgeworth expansion of $f_{Y_n}(x)$ when this density exists, otherwise it is a smooth approximation of the (discrete) probabilities for Y_n.

A corresponding expansion in a series of powers of $1/\sqrt{n}$ can be made for the distribution function. We simply integrate term by term. A useful result is that

$$\int \phi(x)H_r(x)dx = -\phi(x)H_{r-1}(x). \qquad (6.23)$$

This is best checked by differentiating both sides and comparing. Since $\phi'(x) = -x\phi(x)$, we find that

$$\frac{d}{dx}(-\phi(x)H_{r-1}(x)) = \phi(x)\left(xH_{r-1}(x) - H'_{r-1}(x)\right).$$

Using (6.20) for the derivative, we continue with

$$\phi(x)\left(xH_{r-1}(x) - (r-1)H_{r-2}(x)\right) = \phi(x)H_r(x),$$

where we have used (6.19). This verifies (6.23). The distribution function can now be written

$$F_{Y_n}(x) = \Phi(x) - \frac{\phi(x)}{\sqrt{n}}H_2(x)\frac{\mu_3}{6\sigma^3}$$
$$- \frac{\phi(x)}{n}\left(\frac{H_3(x)}{24}\left(\frac{\mu_4}{\sigma^4} - 3\right) + H_5(x)\frac{\mu_3^2}{72\sigma^6}\right) + O\left(\frac{1}{n\sqrt{n}}\right).$$
$$(6.24)$$

6.5.1 Empirical Edgeworth expansion

When we are concerned with variables with an unknown distribution we cannot use the Edgeworth expansion as it stands, since we have no moments of the variables. The natural thing to do is then to find estimates of the moments $\hat{\mu}$, $\hat{\sigma}^2$, $\hat{\mu}_3$, $\hat{\mu}_4$ and use them in the expansion above. In parametric situations we may for instance take maximum likelihood estimated parameters of the distribution and then compute the corresponding moments for this distribution. In non-parametric situations we may use $\bar{X}$, s^2, $\Sigma(X_i - \bar{X})^3/n$, $\Sigma(X_i - \bar{X})^4/n$. Since the higher estimated moments can be very unstable, it can be wise to stop after one term of the Edgeworth expansion in the non-parametric case.

6.5.2 Percentile approximation

We are often interested in approximate confidence intervals, and for this we need percentiles in the distribution of some function of the data and the parameter. Suppose this distribution has an Edgeworth expansion (or some other expansion of this type). One way to find an approximate α-percentile is of course to neglect the ordo term and use some simple numerical method to find the percentile of the Edgeworth approximation. The starting value is typically found from the asymptotic distribution (usually the standard normal distribution where percentiles are easily found). Let $z_0 = z^{(\alpha)}$ denote the percentile of this distribution. If $\hat{F}(z_0)$ is the value of the Edgeworth approximated distribution function at this point, and $\hat{f}(z_0)$ the density approximation, then $z_1 = z_0 - (\hat{F}(z_0) - \alpha)/\hat{f}(z_0)$ is often good enough, and can be iterated. However, convergence is not guaranteed by this method.

A more analytical method is to expand the percentile as a series in $1/\sqrt{n}$. Write this as $z = z_0 + a_1/\sqrt{n} + a_2/n + o(1/n)$. Then plug this argument into the Edgeworth expansion of the distribution function, and series expand the whole thing around z_0. The constants a_1, a_2 are found by taking the result to be $\alpha + o(1/n)$, which means that the $1/\sqrt{n}$ and $1/n$ terms are equated to zero.

From (6.24) we have

$$
F_{Y_n}(z) = \Phi(z) - \frac{1}{\sqrt{n}}\phi(z)H_2(z)\frac{\mu_3}{6\sigma^3}
$$
$$
- \frac{1}{n}\phi(z)\Big(\frac{H_3(z)}{24}\big(\frac{\mu_4}{\sigma^4} - 3\big) + H_5(z)\frac{\mu_3^2}{72\sigma^6}\Big) + O\big(\frac{1}{n\sqrt{n}}\big).
$$

Now the first two terms of a series expansion are

$$
\Phi(z) = \Phi(z_0) + \Big(\frac{a_1}{\sqrt{n}} + \frac{a_2}{n}\Big)\phi(z_0) + \frac{1}{2}\Big(\frac{a_1}{\sqrt{n}} + \frac{a_2}{n}\Big)^2\phi'(z_0)
$$

and

$$\phi(z)H_r(z) = \phi(z_0)H_r(z_0)$$
$$+ (\frac{a_1}{\sqrt{n}} + \frac{a_2}{n})(-z_0\phi(z_0)H_r(z_0) + \phi(z_0)rH_{r-1}(z_0)),$$

where we use (6.20) for H'_r. This gives

$$F_{Y_n}(z) = \Phi(z_0) + \frac{1}{\sqrt{n}}\left[a_1\phi(z_0) - \phi(z_0)H_2(z_0)\frac{\mu_3}{6\sigma^3}\right] +$$
$$\frac{1}{n}\left[a_2\phi(z_0) + \frac{a_1^2\phi'(z_0)}{2} - \frac{a_1(\phi'(z_0)H_2(z_0) + \phi(z_0)H'_2(z_0))\mu_3}{6\sigma^3}\right.$$
$$\left. - \frac{\phi(z_0)H_3(z_0)}{24}(\frac{\mu_4}{\sigma^4} - 3) - \phi(z_0)H_5(z_0)\frac{\mu_3^2}{72\sigma^6}\right] + O(\frac{1}{n\sqrt{n}}).$$

Since $\Phi(z_0) = \alpha$ is the target value, this will be correct up to $o(1/n)$ if we set the brackets to zero, and solve

$$a_1 = \frac{\mu_3}{6\sigma^3}H_2(z_0)$$
$$a_2 = \frac{H_3(z_0)}{24}(\frac{\mu_4}{\sigma^4} - 3) + H_5(z_0)\frac{\mu_3^2}{72\sigma^6} - \frac{\mu_3^2 z_0}{72\sigma^6}H_2^2(z_0)$$
$$+ \frac{\mu_3^2 z_0}{18\sigma^6}H_2(z_0).$$

$$(6.25)$$

We have used $\phi'(z) = -z\phi(z)$, $H'_n(z) = nH_{n-1}(z)$, and $H_1(z) = z$ to simplify the expression for a_2 somewhat. If a first order expansion is wanted, cancel a_2 and use a_1 as above.

Edgeworth expansion can be useful both for the direct adjustment of confidence intervals or their confidence levels and for theoretical studies of various confidence interval methods. It provides for instance a background for the asymptotics of Beran's method below.

For further reading on asymptotic expansion and approximation in general, we refer to Barndorff-Nielsen and Cox (1989). The monograph by Hall (1992) gives the state of the art of Edgeworth expansion in bootstrap analysis.

6.6 Confidence interval methods

Two types of confidence intervals were introduced in Chapter 5. The first and simple interval was based on the asymptotic analogy between the distributions of $\hat{\theta}^* - \tilde{\theta}$ and $\hat{\theta} - \theta$. The second method used studentized expressions where instead the bootstrap distribution of $(\hat{\theta}^* - \tilde{\theta})/\hat{\sigma}^*(\hat{\theta})$ defined the limits of $(\hat{\theta} - \theta)/\hat{\sigma}(\hat{\theta})$. We will here limit the exposition in the first section to an illustration of the studentized interval together with some ingredients of general bootstrap methodology.

We then introduce transformation methods, a double bootstrap method, and a level correction method. Some further versions have been suggested in the voluminous literature but are not covered here. The Bayesian method by Rubin (1981) provides for example an interesting alternative with a different kind of resampling.

6.6.1 A parametric case with studentized interval

The situation considered is from the field of reliability and life time distributions. A sample of observed life times is given as

$$\begin{array}{llll}
2.472 & 0.983\text{D} \; 1.339\text{D} \; 1.435 & 1.229\text{D} \; 3.110 \\
4.711 & 0.413\text{D} \; 6.113 & 0.115\text{D} \; 0.392\text{D} \; 2.600\text{D} \\
2.800\text{D} \; 3.002 & 0.865 & 0.700\text{D} \; 1.456
\end{array}$$

(time unit 1000 hours).

The sign D indicates failure time or death, and the rest are censoring times without failures or deaths on the studied components. We will here think about failures on a technical system. Let x_i, $i = 1, \ldots, n_x$ denote the failure times, and y_j, $j = 1, \ldots, n_y$ the censoring times. Suppose the censoring is due to failures on other parts of the studied system. It is then natural to use a competing risk model for the data. Let the Weibull distribution be postulated for the time to failure if no censoring interferes

$$f(x; a, b) = abx^{b-1}e^{-ax^b}, \quad 0 < x < \infty,$$

with parameters $a > 0$, $b > 0$. Correspondingly the time to censoring, given an infinitely late failure time, is postulated to have another Weibull distribution with positive parameters c and d

$$g(y; c, d) = cdy^{d-1}e^{-cy^d}, \quad 0 < y < \infty.$$

The survival probabilities in the two distributions become $\bar{F}(x; a, b) = 1 - F(x; a, b) = e^{-ax^b}$ and $\bar{G}(y; c, d) = e^{-cy^d}$. Assuming independence, we get the likelihood

$$\begin{aligned} L &= \prod_i f(x_i; a, b)\bar{G}(x_i; c, d) \prod_j \bar{F}(y_j; a, b)g(y_j; c, d) \\ &= \prod_i f(x_i; a, b) \prod_j \bar{F}(y_j; a, b) \prod_i \bar{G}(x_i; c, d) \prod_j g(y_j; c, d) \\ &= L_1(a, b) \, L_2(c, d). \end{aligned} \tag{6.26}$$

The likelihood factorizes, and we can maximize each part on its own. The model for the censoring will thus not affect the likelihood estimates of a and b, only the censoring times do. For the given data we find that $L_1(a, b)$ is maximized for $\hat{a} = 0.299$, $\hat{b} = 0.882$, and $L_2(c, d)$ for $\hat{c} = 0.0624$, $\hat{d} = 2.121$.

The computations behind this result are fairly standard but will be given here for completeness.

$$\ln L_1(a, b) = \sum_{i=1}^{n_x}(\ln a + \ln b + (b-1)\ln x_i - ax_i^b) - \sum_{j=1}^{n_y} ay_j^b$$

$$\frac{\partial \ln L_1}{\partial a} = \sum_i (\frac{1}{a} - x_i^b) - \sum_j y_j^b = 0$$

$$a = \frac{n_x}{\sum x_i^b + \sum y_j^b}.$$

Using this expression for a we can write the next derivative as

$$\frac{1}{n_x} \frac{\partial \ln L_1}{\partial b} = \frac{1}{b} + \frac{\sum \ln x_i}{n_x} - \frac{\sum x_i^b \ln x_i + \sum y_j^b \ln y_j}{\sum x_i^b + \sum y_j^b} = 0.$$

$$(6.27)$$

The left member is a decreasing function of b for $b > 0$. It diverges to $+\infty$ when $b \to 0^+$ and converges to $\sum \ln x_i / n_x - \max_{i,j}(\ln x_i, \ln y_j)$ when $b \to +\infty$. The last limit is negative if at least two different x_i-values exist in the data. Under this mild condition, we have a unique solution $\hat{b}$ which is easily found numerically. The estimates $\hat{c}$ and $\hat{d}$ are found in exactly the same way by interchanging the roles of x_i and y_j.

So far we have made a classical estimation. Now it is time to discuss the parametric bootstrap. Suppose the parameter of interest is

$$\theta = \bar{F}(1.2; a, b) = e^{-a 1.2^b}.$$

This is the probability that a component can survive 1200 hours. Inserting $\hat{a} = 0.299$, $\hat{b} = 0.882$ we get the estimate $\hat{\theta} = 0.704$.

For resampling we interpret the maximum likelihood estimated distribution as closest to the given data. In the parametric bootstrap we therefore draw failure times and censoring times at random from

$$\tilde{f}(x) = f(x; \tilde{a}, \tilde{b}); \quad \tilde{g}(y) = g(y; \tilde{c}, \tilde{d})$$

with parameters $(\tilde{a}, \tilde{b}, \tilde{c}, \tilde{d}) = (\hat{a}, \hat{b}, \hat{c}, \hat{d})$. One consequence is that the (true) bootstrap parameter becomes $\tilde{\theta} = 0.704 = \hat{\theta}$. The resampling can be done as follows.

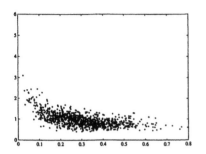 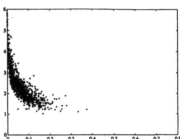

Figure 6.1 *1000 bootstrap replicates of* $(\hat{a}^*, \hat{b}^*)$ *left, and* $(\hat{c}^*, \hat{d}^*)$ *right.*

Let U_k^* and V_k^*, $1 \leq k \leq n_x + n_y$, be random variables drawn fron $\tilde{f}(x)$ and $\tilde{g}(y)$ respectively. If $U_k^* \leq V_k^*$ then let U_k^* be a failure time, and if $V_k^* < U_k^*$ let V_k^* be a censoring time. Collecting failures and censorings we get the bootstrap resample X_i^*, $i = 1, \ldots, n_x^*$ and Y_j^*, $j = 1, \ldots, n_y^*$.

We can generate Weibull distributed data U_k^* as the solution to the equation $1 - \tilde{F}(U_k^*) = e^{-\tilde{a}(U_k^*)^{\tilde{b}}} = R_k$, where R_k is a pseudo-random number uniform in $(0,1)$. This gives $U_k^* = (-\ln(R_k)/\tilde{a})^{(1/\tilde{b})}$.

From the bootstrap data X_i^* and Y_j^* we estimate $\hat{a}^*$, $\hat{b}^*$, $\hat{c}^*$, $\hat{d}^*$ by maximum likelihood just as for the original data, and compute $\hat{\theta}^*$. The result of 1000 bootstrap simulations is illustrated in Figure 6.1.

For a studentized confidence interval we also need an estimate $\hat{\sigma}(\hat{\theta})$ of the standard deviation of $\hat{\theta}$, and a corresponding estimate for each bootstrap sample. The confidence interval will then be based on the bootstrap distribution of $(\hat{\theta}^* - \tilde{\theta})/\hat{\sigma}(\hat{\theta}^*)$. We use asymptotic theory for the estimated standard deviation.

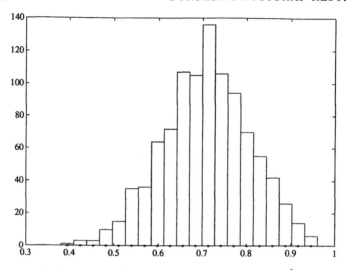

Figure 6.2 *Histogram of 1000 bootstrap replicates of $\hat{\theta}^*$*.

The inverse information matrix can be used as a covariance matrix for $\sqrt{n}(\hat{a}, \hat{b}, \hat{c}, \hat{d})$ as demonstrated for example by Lehmann (1983, Section 6.4). For a variable X, with a sufficiently differentiable density function $f(x;\eta)$ with respect to the parameter vector η, the information matrix is defined as the matrix

$$
\begin{aligned}
\mathbf{I} &= E\left[\left(\frac{\partial \ln f(X;\eta)}{\partial \eta}\right)\left(\frac{\partial \ln f(X;\eta)}{\partial \eta}\right)'\right] \\
&= -E\left[\frac{\partial^2 \ln f(X;\eta)}{\partial \eta \partial \eta'}\right]
\end{aligned}
\tag{6.28}
$$

with obvious modifications for other types of data. As an approximation (estimate) of $\mathbf{I}$ it is sometimes useful to replace the expected value by the corresponding average over data to get

$$
\hat{\mathbf{I}} = -\frac{1}{n}\sum \frac{\partial^2 \ln f(x_i,\eta)}{\partial \eta \partial \eta'} = -\frac{1}{n}\frac{\partial^2 \ln L}{\partial \eta \partial \eta'}.
\tag{6.29}
$$

This generalizes to data with the competing risk structure.

Let **I** be the information matrix for the model of our data, and $\eta = (a, b, c, d)'$. Let $\hat{\mathbf{I}}$ be given by the last member of (6.29) with the likelihood L given by (6.26), and $n = n_x + n_y$. Since the likelihood factorizes, the log likelihood becomes $\ln L = \ln L_1(a, b) + \ln L_2(c, d)$. This produces some zeros in the second derivatives and we get

$$
\hat{\mathbf{I}} = -\frac{1}{n}
\begin{pmatrix}
\frac{\partial^2 \ln L}{\partial a \partial a} & \frac{\partial^2 \ln L}{\partial a \partial b} & 0 & 0 \\
\frac{\partial^2 \ln L}{\partial a \partial b} & \frac{\partial^2 \ln L}{\partial b \partial b} & 0 & 0 \\
0 & 0 & \frac{\partial^2 \ln L}{\partial c \partial c} & \frac{\partial^2 \ln L}{\partial c \partial d} \\
0 & 0 & \frac{\partial^2 \ln L}{\partial c \partial d} & \frac{\partial^2 \ln L}{\partial d \partial d}
\end{pmatrix}
$$

computed at the estimated parameter point.

By a first order series expansion of $\hat{\theta} = \theta(\hat{a}, \hat{b})$, the variances/covariances of $(\hat{a}, \hat{b})$ are transformed into the variance for $\hat{\theta} = \exp(-\hat{a}\, 1.2^{\hat{b}})$. This gives the estimate

$$
\hat{\sigma}^2(\hat{\theta}) = \frac{1}{n}\left(\frac{\partial \theta}{\partial a}, \frac{\partial \theta}{\partial b}, 0, 0\right) \hat{\mathbf{I}}^{-1} \left(\frac{\partial \theta}{\partial a}, \frac{\partial \theta}{\partial b}, 0, 0\right)' \tag{6.30}
$$

computed for (a, b, c, d) at their estimated values. For the given data we get $\hat{\sigma}(\hat{\theta}) = 0.0843$ by this computation. A bootstrap analysis of the pivot $\frac{\hat{\theta} - \theta}{\hat{\sigma}(\hat{\theta})}$ is illustrated in Figure 6.3. This analysis gave the 2.5% and 97.5% percentiles at -2.376 and $+2.365$. This is a lot wider than the ± 1.96 given by a normal approximation. In exact numbers, we end up with $-2.376 < (\hat{\theta} - \theta)/\hat{\sigma}(\hat{\theta}) < 2.365$ or $0.505 < \theta < 0.904$ with 95% confidence.

Remark
By this resampling method, the number n_x^* of observed failures becomes random with a binomial distribution. Also $n_y^* = 17 - n_x^*$ is of course binomial. The special cases $n_x^* < 2$ or $n_y^* < 2$ cause trouble, since the parameters will not be estimable. These cases must therefore have negligible probability in this procedure. If this is not the case, a bootstrap

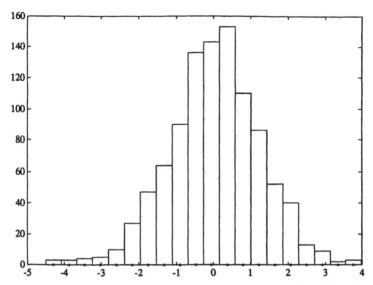

Figure 6.3 *Histogram of 1000 bootstrap replicates of* $(\hat{\theta}^* - \tilde{\theta})/\hat{\sigma}(\hat{\theta}^*)$.

can be defined conditional on having n_x^* in a given set, such as $\{2 \leq n_x^* \leq 15\}$, although the theoretical effects of this conditioning have not been carefully studied.

6.6.2 Transformation based methods

In a series of papers Efron (1979, 1982, 1985, 1987) developed some confidence interval methods for the bootstrap. In the end these were based on fairly refined transformation methods. This development was studied and discussed by other researchers such as Hinkley and Wei (1984), who compared bootstrap and jackknife methods, Hall (1986, 1988) who compared the asymptotic properties of different procedures, DiCiccio and Romano (1988) who surveyed and contributed to the development, and many others. Strangely enough, Efron defined his intervals 'backwards' compared to a strict analogy between $(\hat{\theta} - \theta)$ and $(\hat{\theta}^* - \tilde{\theta})$ or their studentized versions. This was heavily critisized by Hall (1988), defended by some given examples by Efron in the discussion of Hall's work and did come out better for quantiles in Falk and Kaufmann (1991). See also the discussion on this point in Hall (1992). Obviously the question has not been settled yet. For elementary use we have in our earlier sections indicated a preference for 'forward' intervals based on studentized quantities where possible. Although simpler solutions usually exist with the same asymptotic order of correctness as the most advanced transformation based interval, the transformation theory shows an interesting line of thinking. One of the strongest arguments for the transformation based intervals is that if the estimate is transformed by a monotone function, then the confidence limits will be scaled by the same transform. This is not fulfilled by the studentized intervals. Another interesting argument (pointed out by a referee) is that when the range of a parameter is a finite interval, and estimates are confined to the same range, Efron's 'backwards' intervals will necessarily be within this range, but this is not true for intervals defined 'forwards'. The estimation of a high correlation is a good example. Also remember that asymptotic correctness can sometimes be traded against simplicity, and the first transformation intervals are very simple indeed.

Let $X_1, \ldots, X_n$ be a sample with density function $f_\theta(x)$, where θ is a scalar parameter. (Generalization to vector parameters is given in Efron (1987).) Let $\hat{\theta} = \hat{\theta}(X_1, \ldots, X_n)$ estimate θ and have the continuous distribution $H_\theta(s)$. Let $X_1^*, \ldots, X_n^*$ be a parametric resample from $f_{\hat{\theta}}(x)$ ($\tilde{\theta} = \hat{\theta}$ here) and $\hat{\theta}^* = \hat{\theta}(X_1^*, \ldots, X_n^*)$ the bootstrap estimate which gets the bootstrap distribution $H_{\hat{\theta}}(s)$. The parametric bootstrap illustrates best the basic idea and allows more theory so we follow Efron and use it here but the final results are easily interpreted for the non-parametric bootstrap as well.

As background to the transformations, we remind the reader that by a monotone transform of $\hat{\theta}$ we may construct a variable with any distribution we want. In particular the transformation $\Phi^{-1}(H_\theta(\hat{\theta}))$ produces a variable with the standard normal distribution $\Phi(y)$. ($P(\Phi^{-1}(H_\theta(\hat{\theta})) \leq y) = P(\hat{\theta} \leq H_\theta^{-1}(\Phi(y))) = H_\theta(H_\theta^{-1}(\Phi(y))) = \Phi(y)$ if H_θ has an inverse, otherwise use the fact that $H_\theta(\hat{\theta})$ has a rectangular distribution over $(0,1)$.) Transformation to the standard normal is therefore nothing peculiar. Furthermore most estimates are asymptotically normal, so there is hope for rather nice transformation functions.

Now, make the much stronger assumption that (approximately) a monotone increasing transform $g(\hat{\theta})$ exists, independent of the true θ, such that

$$g(\hat{\theta}) - g(\theta) \text{ is } N(0,1) \text{ for any } \theta.$$

This implies that $g(\hat{\theta}^*) - g(\hat{\theta})$ is also $N(0,1)$ in the bootstrap simulations.

Let $z^{(\alpha)}$ be the α-percentile in the $N(0,1)$-distribution. We have

$$\alpha = P^*(g(\hat{\theta}^*) - g(\hat{\theta}) \leq z^{(\alpha)}) = P^*(\hat{\theta}^* \leq g^{-1}(g(\hat{\theta}) + z^{(\alpha)})).$$

Since the bootstrap distribution of $\hat{\theta}^*$ is observed by us, we can identify the percentiles of this distribution. Let ξ_α be the observed α-percentile of $\hat{\theta}^*$. Without knowing the transformation g we have that

$$g^{-1}(g(\hat{\theta}) + z^{(\alpha)}) = \xi_\alpha, \quad \text{the } \alpha\text{-percentile of } \hat{\theta}^*. \quad (6.31)$$

For the original problem we have $\alpha = P(g(\hat{\theta}) - g(\theta) \leq z^{(\alpha)})$. Solving for θ this implies that

$$\alpha = P(g^{-1}(g(\hat{\theta}) - z^{(\alpha)}) \leq \theta).$$

Use the symmetry of $N(0,1)$ which gives $z^{(\alpha)} = -z^{(1-\alpha)}$. This means that

$$\alpha = P(g^{-1}(g(\hat{\theta}) + z^{(1-\alpha)}) \leq \theta)$$

where $g^{-1}(g(\hat{\theta}) + z^{(1-\alpha)}) = \xi_{1-\alpha}$ is the $(1 - \alpha)$-percentile of the bootstrap distribution according to (6.31). A one-sided confidence interval for θ with confidence level $1 - \alpha$ is therefore given as $\theta \le g^{-1}(g(\hat{\theta}) + z^{(1-\alpha)}) = \xi_{1-\alpha}$, and a two-sided with level $1 - 2\alpha$ is given accordingly as $\xi_\alpha < \theta < \xi_{1-\alpha}$. By this so-called *percentile method* we read off the interval directly in the bootstrap distribution of $\hat{\theta}^*$.

The choice of the normal distribution is arbitrary. All we need here and in the following is a distribution allowing us to replace $z^{(\alpha)}$ by $-z^{(\beta)}$ for a known value β.

A good transformation g will not exist in most problems, but the idea can be generalized. The *bias corrected percentile method* BC assumes instead that a transformation g and a constant z_0 exists such that for all θ

$$g(\hat{\theta}) - g(\theta) + z_0 \text{ is } N(0,1).$$

The method can be seen as a special case of the next generalization.

The *accelerated bias corrected percentile method* BCa generalizes to non-constant variance. Now many situations will approximately have the suggested transformations over the parameter range of interest. The underlying assumption is that there exist constants a and z_0 and a monotone increasing transformation g such that

$$\frac{g(\hat{\theta}) - g(\theta)}{1 + ag(\theta)} + z_0 \text{ is } N(0,1) \qquad (6.32)$$

for all θ. Here $1 + ag(\theta)$ represents the standard deviation and must be positive. For $a = 0$ we have the BC method and if also $z_0 = 0$ we get the percentile method. How z_0 and a are determined will be discussed later. If they are determined, we can find the confidence interval for θ without knowing the transformation $g(.)$. The existence is enough.

Start with the observable bootstrap distribution for $\hat{\theta}^*$ and let ξ_α denote the percentile where $P^*(\hat{\theta}^* \le \xi_\alpha) = \alpha$. In terms of the distribution function for $\hat{\theta}^*$ we have $\xi_\alpha = H_{\hat{\theta}}^{-1}(\alpha)$. The same percentile can also be represented via the transformation. Since

$$U^* = \frac{g(\hat{\theta}^*) - g(\hat{\theta})}{1 + ag(\hat{\theta})} + z_0 \text{ is } N(0,1)$$

we have $\alpha = P^*(U^* \leq z^{(\alpha)})$ and also

$$\alpha = P^*\left(\hat{\theta}^* \leq g^{-1}(g(\hat{\theta}) + (z^{(\alpha)} - z_0)(1 + ag(\hat{\theta})))\right).$$

We can therefore identify

$$g^{-1}(g(\hat{\theta}) + (z^{(\alpha)} - z_0)(1 + ag(\hat{\theta}))) = \xi_\alpha, \qquad (6.33)$$

a known value.

A one-sided interval for θ with confidence level $1 - \alpha$ can now be derived from

$$1 - \alpha = P(U > z^{(\alpha)}) = P\left(\frac{g(\hat{\theta}) - g(\theta)}{1 + ag(\theta)} + z_0 > z^{(\alpha)}\right)$$

$$= P\left(\theta < g^{-1}(g(\hat{\theta}) + (1 + ag(\hat{\theta}))\frac{z_0 - z^{(\alpha)}}{1 - a(z_0 - z^{(\alpha)})})\right). \ (6.34)$$

Of course $\hat{\theta}$ is the random quantity. The inequality produces a confidence interval for θ if we can identify the expression. The upper boundary will be of the form (6.33) if we can find β such that $(z_0 - z^{(\alpha)})/(1 - a(z_0 - z^{(\alpha)})) = z^{(\beta)} - z_0$. Then $\theta < \xi_\beta$ will be the resulting confidence interval. But

$$z^{(\beta)} = z_0 + \frac{z_0 - z^{(\alpha)}}{1 - a(z_0 - z^{(\alpha)})}$$

corresponds to

$$\beta = \Phi\Big(z_0 + \frac{z_0 - z^{(\alpha)}}{1 - a(z_0 - z^{(\alpha)})}\Big) = \Phi\Big(z_0 + \frac{z_0 + z^{(1-\alpha)}}{1 - a(z_0 + z^{(1-\alpha)})}\Big),$$
$$(6.35)$$

where the substitution $z^{(\alpha)} = -z^{(1-\alpha)}$ connects the formula to the level $1 - \alpha$ of the interval. The recipe is therefore to compute β from (6.35) and then search the β-percentile ξ_β in the bootstrap simulations. This can be done in both the parametric and the non-parametric setting.

For a lower boundary of θ we replace $1 - \alpha$ by α in the formula for β. A two-sided interval with confidence level $1 - 2\alpha$ is then found from the percentiles $\xi_{\beta_1}, \xi_{\beta_2}$ where $\beta_1 = \Phi\Big(z_0 + \frac{z_0 + z^{(\alpha)}}{1 - a(z_0 + z^{(\alpha)})}\Big)$, and $\beta_2 = \Phi\Big(z_0 + \frac{z_0 + z^{(1-\alpha)}}{1 - a(z_0 + z^{(1-\alpha)})}\Big)$. In terms of the distribution $H_{\hat{\theta}}(.)$ of $\hat{\theta}^*$ this interval is $H_{\hat{\theta}}^{-1}(\beta_1) < \theta < H_{\hat{\theta}}^{-1}(\beta_2)$.

It remains to set the bias parameter z_0 and the acceleration parameter a. The bias is the easy one. Since $g(.)$ is monotone increasing, the bootstrap step gives that

$$P^*(\hat{\theta}^* \le \hat{\theta}) = P^*\Big(\frac{g(\hat{\theta}^*) - g(\hat{\theta})}{1 + ag(\hat{\theta})} + z_0 \le z_0\Big) = \Phi(z_0),$$

since (6.32) applies to U^*. The first term is the proportion of bootstrap values falling below $\hat{\theta}$, and written compactly we have

$$z_0 = \Phi^{-1}(H_{\hat{\theta}}(\hat{\theta})). \qquad (6.36)$$

The acceleration parameter is more involved and related to skewness. Efron (1987) motivates the value

$$a = \frac{1}{6}E\Big[\Big(\frac{\psi - \mu_\psi}{\sigma_\psi}\Big)^3\Big], \qquad (6.37)$$

where $\psi = \frac{\partial \ln h_\theta(s)}{\partial \theta}$ evaluated at $s = \hat\theta$, $h_\theta(s) = \frac{d}{d\theta}H_\theta(s)$ is the probability density of $\hat\theta$, and μ_ψ, σ_ψ are the mean and standard deviation for the random variable ψ. Fortunately, he also suggests a simpler non-parametric value which depends on writing $\theta = \theta(F)$ and $\hat\theta = \theta(F_n)$ as functions of distributions. Then

$$a = \frac{1}{6}\frac{\sum_{i=1}^n \psi_i^3}{(\sum_{i=1}^n \psi_i^2)^{3/2}} \qquad (6.38)$$

where

$$\psi_i = \lim_{\Delta \to 0}\frac{\hat\theta((1-\Delta)F_n + \Delta\delta_i) - \hat\theta(F_n)}{\Delta}$$

is an empirical influence measure, and δ_i is a unit mass of probability at observation x_i (or rather a distribution function rising from 0 to 1 at that value). This acceleration value is best approximated by finite differences for a small Δ in most cases.

Under some fairly strong assumptions Efron (1987) shows second order correctness of parametric intervals based on (6.37) or the equivalent up to $O(1/n)$. This was confirmed and generalized by Hall (1988), although Hall prefers the likewise second order correct studentized method. DiCiccio and Tibshirani (1987) found an explicit transform $g(.)$ with the necessary properties and gave a less computer demanding solution called BC_a^0 to the BCa problem.

Example 6.3
Four different non-parametric intervals, the simple, the percentile, the studentized, and the BCa will be computed for the parameter $\theta = P(X > 2 | X > 1)$ and the following 100 data values sorted by their size. The intervals are two-sided and of nominal 90% confidence level.

Data multiplied by 100

```
  6  11  17  17  19  33  33  36  38  39  42  44
 47  50  53  56  57  57  59  61  62  63  66  67
 67  73  75  75  75  76  79  81  82  83  84  86
 89  98  99 101 104 105 108 109 109 109 110 111
112 112 114 115 116 116 117 125 128 134 134 134
135 136 140 140 143 145 146 147 149 149 155 156
157 158 158 161 161 165 167 168 168 170 174 179
185 189 191 194 197 204 211 212 217 219 241 243
248 262 269 271
```

These data are not real, but simulated from a Weibull distribution with $F(x) = 1 - \exp(-x^2/2)$, $x > 0$, and the true parameter θ is therefore known to be 0.223. Define $Y = \#\{X_i > 1\}$, $Z = \#\{X_i > 2\}$, and $\hat{\theta} = Z/Y$. Observed values are $y = 61$, $z = 11$. Given Y, the variable Z is $Bi(Y, \theta)$ and $\hat{\theta}$ gets the conditional variance $\theta(1-\theta)/Y$. This motivates the simple estimate $\hat{\sigma}(\hat{\theta}) = \sqrt{\hat{\theta}(1 - \hat{\theta})/Y}$. A more complicated unconditional variance can also be used in the studentized interval, but this version is not included here. Our sample gives $\hat{\theta} = \tilde{\theta} = 11/61 = 0.18$, and $\hat{\sigma}(\hat{\theta}) = 0.049$.

We generate 1000 bootstrap samples $X_1^*, \ldots, X_{100}^*$, and compute $\hat{\theta}^*$ and $\hat{\sigma}^*(\hat{\theta})$ for each of them. Let $\hat{\theta}_{(1)}^* \leq \hat{\theta}_{(2)}^*, \cdots \leq \hat{\theta}_{(1000)}^*$ be the ordered results.

In the (90%) inequality $\hat{\theta}_{(50)}^* - \tilde{\theta} < \hat{\theta}^* - \tilde{\theta} < \hat{\theta}_{(951)}^* - \tilde{\theta}$ we replace $\hat{\theta}^* - \tilde{\theta}$ with $\hat{\theta} - \theta$ and get the simple interval (since $\tilde{\theta} = \hat{\theta}$)

$$2\hat{\theta} - \hat{\theta}_{(951)}^* < \theta < 2\hat{\theta} - \hat{\theta}_{(50)}^*.$$

Even simpler is Efron's percentile interval, which gives

$$\hat{\theta}^*_{(50)} < \theta < \hat{\theta}^*_{(951)}.$$

Let $t_{(1)} \leq t_{(2)} \ldots \leq t_{(1000)}$ be the ordered values of $(\hat{\theta}^* - \tilde{\theta})/\hat{\sigma}^*(\hat{\theta})$. The studentized interval is given as

$$\hat{\theta} - t_{(951)}\hat{\sigma}(\hat{\theta}) < \theta < \hat{\theta} - t_{(50)}\hat{\sigma}(\hat{\theta}).$$

The BCa-interval requires z_0 and a. Using that $\hat{\theta}^* < \hat{\theta}$ in 529 of the 1000 simulations we solve $z_0 = 0.0727$ from $\Phi(z_0) = 0.529$. The acceleration parameter is computed from (6.38) but is most easily discussed in terms of Y, Z, and the unit mass δ_i. If $x_i \leq 1$ the mass δ_i is uncounted in both Y and Z. Moving to larger values x_i, we first count the mass in Y and then in both Y and Z. This gives

$$\psi_i = \lim_{\Delta \to 0} \begin{cases} \Delta^{-1}\left(\frac{Z(1-\Delta)}{Y(1-\Delta)} - \frac{Z}{Y}\right) = 0, & x_i = \leq 1; \\ \Delta^{-1}\left(\frac{Z(1-\Delta)}{Y(1-\Delta)+\Delta} - \frac{Z}{Y}\right) = -\frac{Z}{Y^2}, & 1 < x_i \leq 2; \\ \Delta^{-1}\left(\frac{Z(1-\Delta)+\Delta}{Y(1-\Delta)+\Delta} - \frac{Z}{Y}\right) = \frac{Y-Z}{Y^2}, & 2 < x_i. \end{cases}$$

Since there are $Y - Z$ and Z observations in the two last categories, we get $a = \frac{1}{6}(Y - 2Z)/\sqrt{YZ(Y - Z)} = 0.0355$ after some simplification. Formula (6.35) with $z^{(1-\alpha)} = -z^{(\alpha)} = 1.645$ gives $\beta_1 = 0.078$, $\beta_2 = 0.971$. This gives the BCa interval

$$\hat{\theta}^*_{(78)} < \theta < \hat{\theta}^*_{(971)}.$$

Numerically we get the intervals

$0.0975 < \theta < 0.2590$	simple
$0.1017 < \theta < 0.2632$	percentile
$0.1075 < \theta < 0.2802$	studentized
$0.1111 < \theta < 0.2794$	BCa.

Example 6.4

The quality of the four intervals in the preceeding example cannot be assessed from the asymptotic properties alone. Instead we can study the finite sample properties by simulation since we have a well defined situation. We therefore generate 10 000 data sets of 100 data values each, drawn from the distribution function $F(x) = 1 - \exp(-x^2/2)$. Each data set is resampled 1000 times and the confidence intervals are computed exactly as in the example above, of course with updated bias and acceleration estimates for each data set. This takes in all 10^9 generated variables, but can be rationalized somewhat since we only need Y and Z and not all the individual values.

For our 10 000 simulations we observe the following error rates

Lower limit   Upper limit

0.0334	0.0831	simple
0.0379	0.0671	percentile
0.0419	0.0327	studentized
0.0439	0.0475	BCa.

The intended level is an error rate of 0.05 in each tail, and the BCa method came closest to the target in this study followed by the studentized interval. The error probabilities have approximately the standard deviation 0.0023 so we may well have ±0.005 from the true value. Furthermore they are dependent since all four intervals are computed for the same data.

6.6.3 A prepivoting method

In this section we will present a method introduced by Beran (1987, 1988, 1990), where the distribution of a bootstrapped quantity is stabilized by a computer intensive method. The whole procedure boils down to a confidence interval construction by double bootstrap, which illustrates very typically the new ways of thinking. We will also give an asymptotic argument by Beran, the way we interpret him, as to why the method improves the convergence to the intended confidence level. First some notation and fundamentals.

Let $\mathbf{x}_n = (x_1, \ldots, x_n)$ be a set of observed data. Let the corresponding variables $\mathbf{X}_n = (X_1, \ldots, X_n)$ be independent with the distribution function $F(x; \eta)$, where η is a parameter vector. Let $\theta = \theta(F)$ be the one-dimensional parameter of interest. Within the model, θ becomes a function $\theta = T(\eta)$ of the parameter η, but the notation $\theta(F)$ allows a definition also for other distributions, in particular for the empirical. For every sample size n, let $\hat{\theta}(\mathbf{x}_n)$ be the estimate of θ, $\hat{\eta}(\mathbf{x}_n)$ the estimate of η. In this notation we suppress the fact that the functions themselves depend on n.

Since we are primarily interested in confidence intervals, we will use what Beran (1987) calls *roots*, defined as functions of both the data and the unknown parameter value θ. Let $R(\mathbf{x}_n, \theta)$ be such a root, and $R(\mathbf{X}_n, \theta)$ the corresponding model version. To be useful, the distribution of the root must be known or estimable without knowing the value of θ. Sometimes this is fulfilled only asymptotically as $n \to \infty$. Confidence intervals can be obtained from the roots by solving $\{\theta; R(\mathbf{x}_n, \theta) \leq c\}$, where c is a value such that $P(R(\mathbf{X}_n, \theta) \leq c) = 1 - \alpha$, the confidence level, exactly or approximately.

Pivot variables are a special kind of root. Since pivots by definition have the same distribution for all values of the parameter, like $R = (\bar{X} - \mu)/(s/\sqrt{n})$ for a sample from the normal distribution, they are of course of high interest. For distributions in general, useful exact pivots may not exist, but studentized estimates like $(\hat{\theta} - \theta)/\hat{\sigma}(\hat{\theta})$, where $\hat{\sigma}(\hat{\theta})$ is a sensible estimator of the standard deviation of $\hat{\theta}$, are often approximate pivots in the sense that the distribution does not vary much when we vary the parameter. For large data sets, the distribution will usually be close to the standard normal.

In bootstrap analysis the conclusions are based on bootstrap samples from a distribution which is typically somewhat different from the true one. It is then a great advantage to resample pivots whose statistical properties remain relatively unaffected by the change of distribution.

A root variable with unknown distribution can be transformed into an approximate pivot variable by a bootstrap method. Defining the new variable as a new root, another outer bootstrap analysis estimates the distribution of this root function. This double bootstrap typically gives a higher order convergence to the limit distribution, as we will prove later for a special situation. From this follows a higher precision in confidence levels computed from the roots, at least for sufficiently large samples sizes, In principle, the method can be iterated to arbitrary order. However, the double bootstrap is already a heavy computational effort and it is hardly of present interest to go any further. The method was introduced by Beran (1987) for a single interval, extended to simultaneous confidence intervals with equal individual levels in Beran (1988, 1990), and the reader is referred to these articles for more details and examples than we can give here. The method goes as follows.

Let $R = R(\mathbf{X}_n, \theta)$ be a root useful for the computation of a confidence interval for θ when the observed $\mathbf{x}_n$ replaces $\mathbf{X}_n$. For an estimate $\hat{\theta} = \hat{\theta}(\mathbf{X}_n)$, we can think of the root $|(\hat{\theta} - \theta)/\hat{\sigma}(\hat{\theta})|$ if a two-sided interval is wanted and we have an estimate of the standard deviation for $\hat{\theta}$.

Let R have the distribution function $G(x; \eta)$ where η is the full parameter. It is well known, and easily checked, that if a variable Y has a continuous distribution $G(y)$, then the transformed variable $Z = G(Y)$ has a rectangular distribution over the interval (0,1), so that $P(Z \leq z) = z$, for $0 < z < 1$. Applied to the root R, here supposed to have a continuous distribution, we have that

$$G(R; \eta) \text{ is } \text{Re}(0, 1). \tag{6.39}$$

The distribution G is typically unknown, but we can use the bootstrap to estimate it. Both non-parametric and parametric resampling are possible depending on the modelling situation. First we have to resample the x-observations from some distribution $\hat{F}(x)$. We use the notation $\hat{F}$ to cover both the non-parametric F_n and parametric versions. Let $\tilde{\theta} = \theta(\hat{F})$ be the value of the parameter in the bootstrap distribution, $R^* = R(\mathbf{X}_n^*, \tilde{\theta})$ the resampled roots, and estimate

$$\hat{G}(x) = \hat{G}(x|\mathbf{x}_n) = P_*(R^* \leq x)$$

as the distribution function of the root in the bootstrap simulations. We can already make a first confidence interval, with intended confidence level $1 - \alpha$, by searching a value c such that $\hat{G}(c) = 1 - \alpha$. From the result

$$P_*(R(\mathbf{X}_n^*, \tilde{\theta}) \leq c) = 1 - \alpha$$

it is then concluded that

$$P(R(\mathbf{X}_n, \theta) \leq c) \approx 1 - \alpha.$$

The first confidence interval for θ is therefore given by substituting the observations $\mathbf{x}_n$ for $\mathbf{X}_n$ and solving

$$I_1 = \{\theta; R(\mathbf{x}_n, \theta) \leq c\}. \tag{6.40}$$

The result of this procedure will be more reliable if $R(\mathbf{X}_n, \theta)$ already is an approximate pivot. The question is how to find one. Looking back at the uniform distribution of $G(R; \eta)$ in (6.39), this transformation has a distribution which is always the same (for continuous variables). For known $G(.; \eta)$, the variable $G(R(\mathbf{X}_n, \theta); \eta)$ is therefore a perfect pivot, but $G(.; \eta)$ is unknown. Instead we have the bootstrap estimate $\hat{G}$ computed from the data, and Beran (1987) uses $\hat{G}(R(\mathbf{X}_n, \theta))$ as an approximate pivot variable. Now the interpretation becomes a little bit tricky. If $\hat{G}$ is held fix, this is not a root variable. The estimate $\hat{G}$ is a function of the original data $\mathbf{x}_n$ and if held fix, $\hat{G}(R(\mathbf{X}_n, \theta))$ would become a function of three arguments $\mathbf{X}_n, \mathbf{x}_n$, and θ. If instead the procedure leading to $\hat{G}$ is imagined as repeated on every out-

come of $\mathbf{X}_n$, then $\hat{G}$ becomes a function of $\mathbf{X}_n$. So defined
we have the new root

$$R_2(\mathbf{X}_n, \theta) = \hat{G}(R(\mathbf{X}_n, \theta)).$$

Let $H(x; \eta)$ be the distribution function of R_2. We now
repeat all the steps as before, with bootstrap roots $R_2^* = R_2(\mathbf{X}_n^*, \tilde{\theta})$, and estimate

$$\hat{H}(x) = P_*(R_2^* \leq x)$$

as the distribution function of the new root in the bootstrap
simulations. Since $\hat{G}$ above was based on a bootstrap analysis, we have to fully repeat this for every run of the R_2^* in the
outer bootstrap. In the end we search a value d such that
$\hat{H}(d) = 1 - \alpha$. From this value we have that

$$P_*(R_2^* \leq d) = 1 - \alpha$$

and infer from this that $P(R_2(\mathbf{X}_n, \theta) \leq d) \approx 1 - \alpha$. The
confidence interval is now defined as

$$
\begin{aligned}
I_2 &= \{\theta; \ R_2(\mathbf{x}_n, \theta) \leq d\} = \{\theta; \hat{G}(R(\mathbf{x}_n, \theta)) \leq d\} \\
&= \{\theta; R(\mathbf{x}_n, \theta) \leq e\},
\end{aligned}
\tag{6.41}
$$

where $\hat{G}(e) = d$ and $\hat{H}(d) = 1 - \alpha$ or more compactly
$\hat{H}(\hat{G}(e)) = 1 - \alpha$. We assume there is a unique such number
e, or take an approximation.

Flow chart of prepivoting double bootstrap
Let $\mathbf{x}_n$, $R(\mathbf{x}_n, \theta)$, $\hat{F}(x)$, $\hat{\theta} = \hat{\theta}(\mathbf{x}_n)$, $\tilde{\theta} = \theta(\hat{F}(.))$ be given.

$i2 = 1, \ldots, B2$	outer bootstrap loop
$\quad \mathbf{X}^*(i2)$	resample from $\mathbf{x}$
$\quad \hat{F}^*$	empirical df of $\mathbf{X}^*(i2)$
$\quad \tilde{\theta}^* = \theta(\hat{F}^*)$	
$\quad R^* = R(\mathbf{X}^*(i2), \tilde{\theta})$	
$\quad i1 = 1, \ldots, B1$	inner bootstrap loop
$\quad\quad \mathbf{X}^{**}(i2, i1)$	resample from $\mathbf{X}^*(i2)$
$\quad\quad R^{**} = R(\mathbf{X}^{**}(i2, i1), \tilde{\theta}^*)$	
$\quad\quad$ next $i1$	
$\quad \hat{G}^*(.)$	df of R^{**} for fixed $i2$
$\quad R_2^* = \hat{G}^*(R(\mathbf{X}^*(i2), \tilde{\theta}))$	
$\quad$ next $i2$	
$\hat{G}(.)$	distribution of R^*
$\hat{H}(.)$	distribution of R_2^*
$d = \hat{H}^{-1}(1 - \alpha)$	
$e = \hat{G}^{-1}(d)$	
$I_2 = \{\theta; R(\mathbf{x}, \theta) \leq e\}$	confidence interval

6.6.4 Asymptotic properties of the prepivoting

Will the prepivoting produce more accurate confidence levels? We will give an asymptotic answer to this fundamental question, following Beran, but we are not able to say anything general about small samples. The articles by Beran referred to above give some interesting finite examples.

Consider a parametric class of distributions and continuous variables. Also consider the parametric bootstrap for simplicity, although the equivalent reasoning can be carried through for the non-parametric version, if we let the jumps of the distributions limit the possible order of convergence.

The necessary notation was introduced in the preceeding section and Section 6.2. Let the sample size $n \to \infty$.

Assumption 1: Let the sequence of estimates $\hat{\eta} = \hat{\eta}(X_n)$ be $\sqrt{n}$-consistent for the full parameter vector η, and let $\hat{\theta}$ be $\sqrt{n}$-consistent for $\theta = \theta(F) = T(\eta)$.

Most estimates fulfil these conditions.

Assumption 2: Let the distribution $G(x; \eta) = G_n(x; \eta)$ of the root R converge to the same non-degenerate distribution $G_0(x)$ for all η, and have the expansion (6.5). Possibly some of the lower order terms are already absent. Let each function of the expansion have bounded x-derivative, and bounded (first) η-derivatives in a neighbourhood of the true η. Let the terms of order r and above sum to $O(n^{-\frac{r}{2}})$ uniformly in x and in the neighbourhood of η. Also let the distribution $H(x; \eta)$ of the second root $R_2 = \hat{G}(R)$ have the same properties when $n \to \infty$.

If the original data are from $F(x; \eta)$, the parametric bootstrap means that we resample from $\hat{F}(x) = F(x; \hat{\eta})$ and have $\tilde{\theta}_n = T(\hat{\eta})$ as the bootstrap parameter. This gives the bootstrapped roots $R^* = R(X_n^*, \tilde{\theta})$. Double bootstrapped roots are constructed similarly with X_n^*, $\tilde{\theta}^*$ replacing the original data and $\tilde{\theta}$.

Statement
Under the assumptions and circumstances above, confidence intervals computed by the double bootstrap with prepivoting have one order $(1/\sqrt{n})$ higher convergence rate than intervals without prepivotion.

Proof The distribution of $R = R(X_n, \theta)$ with $\theta = T(\eta)$ can be written

$$P(R \leq x) = G_n(x; \eta) = G_0(x) + n^{-\frac{r}{2}} g_{nr}(x; \eta) + O(n^{-\frac{r+1}{2}}) \tag{6.42}$$

for some integer $r \geq 1$. First we compute the precision of the single bootstrap. Since the distribution of R^*, i.e. $\hat{G}(x) =$

$G_n(x; \hat{\eta})$, will be used instead of $G(x) = G_n(x; \eta)$ to judge the confidence level, we compare the two distributions using (6.42).

$$
\begin{aligned}
\hat{G}(x) &= G_0(x) + n^{-\frac{r}{2}} g_{nr}(x; \hat{\eta}) + O(n^{-\frac{r+1}{2}}) \\
&= G_0(x) + n^{-\frac{r}{2}} g_{nr}(x; \eta) \\
&\quad + n^{-\frac{r}{2}}(g_{nr}(x; \hat{\eta}) - g_{nr}(x; \eta)) + O(n^{-\frac{r+1}{2}}) \\
&= G(x) + n^{-\frac{r}{2}}(\hat{\eta} - \eta)\frac{\partial g_{nr}}{\partial \eta}(\xi) + O(n^{-\frac{r+1}{2}}) \quad (6.43)
\end{aligned}
$$

for some ξ between $\hat{\eta}$ and η. The $\sqrt{n}$-consistency gives that with probability $1-\varepsilon$, $\|\hat{\eta}_n - \eta\| < M/\sqrt{n}$, $n > n_0$ for some M and n_0. Since $\frac{\partial g}{\partial \eta}$ is bounded, the difference between $\hat{G}$ and G is, with the same probability, absorbed into $O(n^{-\frac{r+1}{2}})$. The conclusion is that $\hat{G}(x) = G(x) + O(n^{-\frac{r+1}{2}})$ with probability $1 - \varepsilon$ or equivalently

$$
\hat{G}(x) = G(x) + O_p(n^{-\frac{r+1}{2}}). \quad (6.44)
$$

This defines the level error of the first interval I_1 as $1 - \alpha + O(n^{-\frac{r+1}{2}}) \pm \varepsilon$, for ε arbitrarily small.

Next we study the second root $R_2 = \hat{G}(R)$. It has a distribution function which we denote $H(x) = H_n(x; \eta)$. Suppose the same regularity conditions are valid as for G. Using (6.43), (6.44) and the uniformity of the approximations we have

$$
H(x) = P(\hat{G}(R) \leq x) = P(G(R) + O(n^{-\frac{r+1}{2}}) \leq x) \pm \varepsilon.
$$

But G is by definition the distribution function of R, and $G(R)$ has the exact $\mathrm{Re}(0,1)$-distribution. For $0 \leq x \leq 1$ we therefore have

$$P(R_2 \leq x) = P(G(R) \leq x + O(n^{-\frac{r+1}{2}})) \pm \varepsilon$$
$$= x + O(n^{-\frac{r+1}{2}}) \pm \varepsilon$$
$$= x + n^{-\frac{r+1}{2}} k(x; \eta) + O(n^{-\frac{r+2}{2}}) \pm \varepsilon. \quad (6.45)$$

Here we have extracted the lowest order term from the Ordo-term. But apart from the ε-term, this is similar to (6.42) with the rectangular distribution replacing the general $G_0(x)$.

Supposing k is differentiable and behaves like g_{nr} above, we can repeat all the steps up to (6.44) with r replaced by $r + 1$. Another arbitrarily small term ε' will add to the first ε. This gives $\hat{H}(x) = H(x) + O_p(n^{-\frac{r+2}{2}})$. The level of $I_2 = \{\theta; R_2 \leq d\}$, where $\hat{H}(d) = 1 - \alpha$, is therefore $1 - \alpha + O(n^{-\frac{r+2}{2}}) \pm \varepsilon''$ within an arbitrarily small ε''. Thus the prepivoting increases the order of the error term by one.

6.6.5   Loh's level adjustment

Suppose we have an approximate test or confidence interval which may be a bit crude. Let α be the intended error probability and $\pi(\alpha)$ the resulting true error probability. Suppose $\pi(\alpha) \neq \alpha$. Now there must in most problems exist a different level α' such that, by the same construction, $\pi(\alpha') = \alpha$. Loh (1987, 1988) has developed a bootstrap method for approximating α' by a value α_1. By this method the precision will typically be of the same asymptotic order as the best alternative bootstrap procedures, provided the first test or confidence interval is asymptotically sound. More precisely Loh shows that starting from the asymptotic normal approximation his method gives the correct $1/\sqrt{n}$-term of the Edgeworth expansion but differs in the $1/n$-term. Thus for one-sided intervals this is second order equivalent to studentized intervals and BCa intervals and this means second order correct (with an error $O(1/n)$) for the intervals studied by Hall (1988). For two-sided intervals Loh shows that his method is one order better than both studentized and BCa

intervals for the same situation. Since the method is simple
and direct, it deserves to be widely used. The method goes
as follows.

Define an unbounded function $\psi(\alpha)$ on $0 < \alpha < 1$, such
that ψ has both continuous third derivative and positive
first derivative. Among possible choices are $\psi(\alpha) = \Phi^{-1}(\alpha)$,
$\psi(\alpha) = \tan(\pi(\alpha - 0.5))$ and $\psi(\alpha) = \ln(\alpha/(1-\alpha))$. Resample
the data and bootstrap analyse the test or confidence inter-
val construction. Let $\hat{\pi}(\alpha)$ be the estimated error level from
this analysis. Now solve α_1 from the equation

$$\psi(\alpha_1) = \psi(\alpha) - (\psi(\hat{\pi}(\alpha)) - \psi(\alpha)) \qquad (6.46)$$

and recompute the original test or confidence interval as if
α_1 were the intended level. The reader is referred to the
articles by Loh for series expansions in some known cases.
Here we will only remark that when $\psi(\alpha)$ is approximately
linear over the region of interest, say $\psi(\alpha) = a + b\alpha$, and
approximately $\hat{\pi}(\alpha) = \pi(\alpha) = \alpha + c$ over the same region we
have by (6.46)

$$\psi(\alpha_1) = a + b\alpha_1 = a + b\alpha - (a + b(\alpha + c) - a - b\alpha)$$

with the solution $\alpha_1 = \alpha - c$ and therefore $\pi(\alpha_1) = \alpha$. This
gives an idea of why things work so nice asymptotically.

Results on bootstrap coverage correction are also given in
Martin (1990).

Example 6.5
Four intervals were computed in Example 6.3 and evaluated
in Example 6.4. If we choose $\psi(\alpha) = \ln(\alpha/(1 - \alpha))$ the
correction (6.46) gives that

$$\ln \frac{\alpha_1}{1 - \alpha_1} = \ln \frac{\alpha}{1 - \alpha} - (\ln \frac{\hat{\pi}(\alpha)}{1 - \hat{\pi}(\alpha)} - \ln \frac{\alpha}{1 - \alpha}).$$

This gives the multiplicative correction

$$\frac{\alpha_1}{1-\alpha_1} = \frac{\alpha}{1-\alpha} \frac{\frac{\alpha}{1-\alpha}}{\frac{\hat{\pi}(\alpha)}{1-\hat{\pi}(\alpha)}}.$$

The simple interval had error probabilities 0.0334 and 0.0839 at the lower and upper limits instead of the nominal $\alpha = 0.05$ at each limit. The corrected values α_1 therefore become 0.074 and 0.029. For the percentile interval we find by the same correction that 0.066 and 0.037 are the new target error probabilities, and for the studentized and the BCa intervals we should aim at the errors 0.060, 0.076 and 0.057, 0.053 respectively.

A new evaluation of the intervals, where we generate 10 000 data sets of 100 data values each and resample each set 1000 times, exactly as in Example 6.4, but with the corrected α-values, gives the following error rates which are considerably closer to 5%.

Lower limit	Upper limit	
0.0554	0.0613	simple
0.0523	0.0511	percentile
0.0531	0.0629	studentized
0.0503	0.0508	BCa.

6.7 Bootstrapping regression models

In this section we will discuss resampling methods in various regression situations. Very little will be said about deeper theoretical questions, asymptotic properties and the like. Freedman (1981), Navidi (1989), and Wu (1986) are recommended reading on such aspects.

Regression has many faces. Some of them were mentioned briefly in Chapter 1. The linear model $Y_i = x_i'\beta + \varepsilon_i$ has been thoroughly analysed, and when ε_i are $N(0,\sigma)$ and independent all the standard problems seem to be well known and have good classical solutions. However, we need not go far to find questions which are not solved so easily. Suppose we want to find the maximum of $\mu(t) = \beta_0 + \beta_1 t + \beta_2 t^2$ and the maximizing t-value when $Y_i = \mu(t_i) + \varepsilon_i$, $i = 1, \ldots, n$ are observed. Now $t_0 = -\beta_1/2\beta_2$, $\mu(t_0) = \beta_0 - \beta_1^2/4\beta_2$, is the extremal point and a maximum if $\beta_2 < 0$. The statistical analysis of the estimated maximum has no direct classical solution. We can of course derive an analysis on the classical basis, but we can also use a bootstrap approach. The situation is extra thrilling if the sign of β_2 is in doubt and if t is bounded to some interval $a \leq t \leq b$.

New questions arise if $Y_i = g(\mathbf{x}_i, \beta) + \varepsilon_i$, has an expected value which is non-linear in β, or if ε_i are not Gaussian or have varying variance. Another challenge is provided by the class of generalized linear models where the parameter θ of some classical parametric model depends on a linear expression as $\theta = h(\mathbf{x}'\beta)$. In such models the usual regression formulation with additive noise is less useful. Maximum likelihood solves many problems here, but others can be analysed by bootstrap methods if we can define a resampling procedure.

The most difficult regression problems come with large predictor sets where model selection is involved. The validation methods of Chapter 3 can be accompanied and sometimes substituted by bootstrap solutions. The outstanding bootstrap contribution is to give uncertainties of the estimated parameters also in very complicated analyses. However, the bootstrap generation of a resample must take into consideration that in model selection problems the model is not well defined.

6.7.1 Basic residual resampling

In the linear regression model $Y_i = \mathbf{x}_i'\beta + \varepsilon_i$ the errors ε_i are independent and identically distributed in the standard case. The most natural resampling strategy is then to try resampling the ε_i. Unfortunately, the errors are not observable, only $(Y_i, \mathbf{x}_i')$ are known to us. The usual solution is then to use the estimate $\hat{\beta} = (\mathbf{X}'\mathbf{X})^{-1}\mathbf{X}'\mathbf{Y}$, where $\mathbf{X} = (\mathbf{x}_1, \ldots, \mathbf{x}_n)'$, $\mathbf{Y} = (Y_1, \ldots, Y_n)'$ and compute residuals $\mathbf{R} = (e_1, \ldots, e_n)'$ as $\mathbf{R} = \mathbf{Y} - \mathbf{X}\hat{\beta}$.

A direct residual resampling gives

$$Y_i^* = \mathbf{x}_i' \hat{\beta} + \varepsilon_i^* \qquad (6.47)$$

where ε_i^* are independent from the empirical distribution of $(e_1, \ldots, e_n)$.

The distribution of e_i is not exactly that of ε_i. We can write $\mathbf{R} = (\mathbf{I} - \mathbf{X}(\mathbf{X}'\mathbf{X})^{-1}\mathbf{X}')\mathbf{Y}$ and find that $E[\mathbf{R}] = \mathbf{0}$, $\mathrm{Cov}(\mathbf{R}) = \sigma^2(\mathbf{I} - \mathbf{X}(\mathbf{X}'\mathbf{X})^{-1}\mathbf{X}')$. Let division and square root be defined elementwise for vectors. Then

$$\mathbf{R}_1 = \mathbf{R}/\sqrt{\mathrm{diag}(\mathbf{I} - \mathbf{X}(\mathbf{X}'\mathbf{X})^{-1}\mathbf{X}')} \qquad (6.48)$$

gives a vector with the same mean and variance as ε_i. A natural alternative is now to resample as above with ε_i^* drawn from the elements of $\mathbf{R}_1$ instead of $\mathbf{R}$. We call this standardized residual resampling.

6.7.2 Vector resampling

A crude but generally available resampling method is to regard the couples $(Y_i, \mathbf{x}_i)$, $1 \leq i \leq n$, as the data units and resample among them. We call this vector resampling. This possibility was considered already in Efron (1982) together with residual resampling. The result of this method gives a different interpretation to the uncertainties. In the traditional analysis we consider the design $\mathbf{X}$ as deterministic, or condition on $\mathbf{X}$ to arrive at the same conclusions. This conditioning is rational since it measures the information we happen to get about the β-values and the amount of information depends directly on $\mathbf{X}$. By the vector resampling, the bootstrap design $\mathbf{X}^*$ gets random, and the estimates will typically show up more variability. However, sometimes vector resampling is the only feasible thing to do. Even if the uncertainties have a tendency to get exaggerated, they can still be useful as conservative estimates of the variances.

In spite of the difficulties with design variation, vector resamplings have some definitive advantages compared to residual resampling. When the error distribution varies with

the design variables (let it be skewness or heteroscedasticity, i.e. varying variance), the residual resampling will hide
this fact, and the same may happen if a non-linearity in
the model is not properly modelled. These phenomena remain unchanged in the vector resampling. For large data sets
without very influential data, vector resampling is therefore
a very attractive procedure when we suspect such effects.
For smaller data sets or data with influential observations
the balance between vector and residual resampling is more
to the advantage of residual resampling.

Some measures can be made to correct for the design varia-
tion. If a linear expression $\mathbf{v}'\beta$ is bootstrapped (where $\mathbf{v}$ is a
deterministic vector) then $\operatorname{Var}(\mathbf{v}'(\hat{\beta} - \beta)) = \sigma^2 \mathbf{v}'(\mathbf{X}'\mathbf{X})^{-1}\mathbf{v}$.
In the bootstrap analysis we may compensate for the design
by considering

$$\frac{\mathbf{v}'(\hat{\beta}^* - \tilde{\beta})}{\sqrt{\mathbf{v}'(\mathbf{X}^{*\prime}\mathbf{X}^*)^{-1}\mathbf{v}}}$$

and apply the result to the corresponding expression for the
original data and estimates. This is not an exact standardization, since repeated use of the same observation gives a
different variance. For non-linear functions of β the natural
thing to do is to linearize by a series expansion around the
estimated point.

6.7.3 Projected residuals

A smooth resampling method can be derived by a geometric
argument. Let $\mathbf{Y} = \mathbf{X}\beta + \varepsilon$ be a vector in the n-dimensional
space R^n and β a vector in R^{p+1}. Define $L(\mathbf{X})$ as the linear subspace of R^n generated by the columns of $\mathbf{X}$, ($L(\mathbf{X})$
contains all vectors of the form $\mathbf{X}\beta$ and has dimension $p+1$
in non-singular cases). Let $M(\mathbf{X})$ be the subspace of R^n orthogonal to $L(\mathbf{X})$, which contains the remaining $n - p - 1$
dimensions. Now, given that the model is correct, the projection of $\mathbf{Y}$ on any (unit) vector of $M(\mathbf{X})$ is exactly the
projection of ε on the same vector since $\mathbf{X}\beta$ is orthogonal

to $M(\mathbf{X})$. Under the standard assumptions that ε has idependent components with mean zero and variance σ^2, the projection of ε on any direction also gets mean zero and variance σ^2. (Use the fact that if $\mathbf{Y}$ has covariance matrix $\mathbf{C}_Y = E[(\mathbf{Y} - \mu_Y)(\mathbf{Y} - \mu_Y)']$, then $\mathbf{AY}$ has the covariance matrix $\mathbf{AC}_Y\mathbf{A}'$. Now $\mathbf{C}_Y = \mathbf{C}_\varepsilon = \sigma^2\mathbf{I}$ since the components are independent with variance σ^2. Let $\mathbf{v}$ be a unit vector. Then $\mathrm{Var}(\mathbf{v}'\varepsilon) = \mathbf{v}'\mathbf{C}_\varepsilon\mathbf{v} = \sigma^2\mathbf{v}'\mathbf{v} = \sigma^2$, and if $\mathbf{v}$ is in $M(\mathbf{X})$ then $\mathbf{v}'\mathbf{Y} = \mathbf{v}'\varepsilon$.) For projections in $M(\mathbf{X})$ we can therefore use the observable vector $\mathbf{Y}$ instead of ε and get the same result.

The projection resampling is defined by drawing independent random unit vectors $\mathbf{v}_i$ in $M(\mathbf{X})$ and letting $\mathbf{Y} = \mathbf{X}\hat{\beta} + \varepsilon^*$, where $\varepsilon^* = \mathbf{VY}$, $\mathbf{V} = (\mathbf{v}_1, \ldots, \mathbf{v}_n)'$. The result is a continuous distribution for bootstrap residuals which is symmetric, and for large n comes close to generating $N(0, s)$-distributed values. The density of the residuals is

$$f(e|\mathbf{Y}) = \frac{\Gamma(\frac{n-p-1}{2})}{\sqrt{\pi}\Gamma(\frac{n-p-2}{2})k}\left(1 - \frac{e^2}{k^2}\right)^{\frac{n-p-4}{2}}, \quad -k < e < k,$$

where $k^2 = (n - p - 1)s^2$.

Complicated as this method may seem, it becomes very simple in a high level computer language such as Matlab or Gauss. In Matlab set rand('normal') to give $N(0,1)$ random numbers, or use the command randn instead of rand in the latest Matlab version. A full vector of n bootstrap residuals is given by the commands

```
U=rand(n,n); U=U−X*(X\U);
norm=sqrt(sum(U.∧2)'); estar=(U'*Y)./norm;
```

Here $X\backslash U$ produces regression coefficients for the U-columns on $\mathbf{X}$, and the second command produces unnormalized random vectors in $M(\mathbf{X})$.

6.7.4 Non-linear regression

Residual resampling and vector resampling can both be applied to non-linear regression of the type

$$Y_i = g(\mathbf{x}_i; \beta) + \varepsilon_i,$$

if the ε_i are equally distributed. In residual resampling we just estimate β, perhaps by least squares, and compute the residuals $e_i = Y_i - g(\mathbf{x}_i; \hat{\beta})$. Each ε_i^* is then drawn at random from $e_1, \ldots, e_n$ before we set $Y_i^* = g(\mathbf{x}_i; \hat{\beta}) + \varepsilon_i^*$.

There is no direct analogue of standardized residuals except when a first order series expansion is adequate and gives a linerized situation of the linear regression type. Efron (1988, Section 5 ff) gives a good illustration of residual resampling in a non-linear situation.

Vector resampling is simply the resampling of n rows of data drawn with replacement from the original n rows. Estimation is done just as for the original data, but as the evaluation of uncertainty has a very general bootstrap solution, we have great freedom to choose the method.

6.7.5 Abstract residual resampling

An elegant confidence interval construction in regression is suggested by Holm (1990). He defines an abstract resampling of the true ε_i and derives observable results although the residuals are not observable themselves. We will interpret his method for a function of the parameter.

Consider the linear model $\mathbf{Y} = \mathbf{X}\beta + \varepsilon$ and suppose we want a two-sided confidence interval with confidence level $c = 1 - 2\alpha$ for some scalar function $g(\beta)$. Write

$$\varepsilon = \mathbf{Y} - \mathbf{X}\beta$$

and draw a resample $\varepsilon^* = (\varepsilon_1^*, \ldots, \varepsilon_n^*)'$, where each $\varepsilon_i^* = \varepsilon_{j(i)}$ and the $j(i)$ are drawn at random from $1, \ldots, n$ independently for different i. This is abstract since ε is not observable, only $\mathbf{Y}$ and $\mathbf{X}$ are. Let $\mathbf{Y}^{\circ} = (Y_{j(1)}, \ldots, Y_{j(n)})'$, $\mathbf{X}^{\circ} = (\mathbf{x}_{j(1)}, \ldots, \mathbf{x}_{j(n)})'$, and

$$\mathbf{Y}^* = \mathbf{X}\beta + \varepsilon^* = \mathbf{X}\beta + (\mathbf{Y}^{\circ} - \mathbf{X}^{\circ}\beta). \tag{6.49}$$

Here the design matrix $\mathbf{X}$ has not changed since $\mathbf{X}^{\circ}$ only affects the residuals. Furthermore in the abstract world, we are free to use the theoretical β although it is unknown. The corresponding bootstrap estimate of β becomes

$$\hat{\beta}^* = (\mathbf{X}'\mathbf{X})^{-1}\mathbf{X}'\mathbf{Y}^* = (\mathbf{X}'\mathbf{X})^{-1}\mathbf{X}'(\mathbf{X}\beta + \mathbf{Y}^{\circ} - \mathbf{X}^{\circ}\beta)$$
$$= \beta + (\mathbf{X}'\mathbf{X})^{-1}\mathbf{X}'\mathbf{Y}^{\circ} - (\mathbf{X}'\mathbf{X})^{-1}\mathbf{X}'\mathbf{X}^{\circ}\beta. \tag{6.50}$$

If we allow β to vary, this becomes a function $\hat{\beta}^*(\beta)$ of β. For the true value of β, the generated $\hat{\beta}^*(\beta)$ have a bootstrap distribution which mimics the distribution of $\hat{\beta}$ itself. This is not true for other β. By the usual confidence interval philosophy, $\hat{\beta}$ ought to be in the central part of its own distribution, and if the bootstrap distribution comes close, then $\hat{\beta}$ should loosely speaking be in the central $c = 1 - 2\alpha$ probability of this distribution with approximately the same probability c. On a scale determined by the function $g(\beta)$ we will accept all β-values such that this central property occurs.

There is one complication, however. If the model has a constant β_0, and $\mathbf{X}$ correspondingly has a constant column (of ones), the parameter β_0 disappears. This is perhaps best seen in (6.49) where $(\mathbf{X} - \mathbf{X}^{\circ})$ gets zeros in the corresponding column. The manipulations that follow are therefore not informative about β_0. (An independent assessment of the variability of β_0 is given by the same component in $(\mathbf{X}'\mathbf{X})^{-1}\mathbf{X}'\mathbf{Y}^{\circ}$ in (6.50).) We will therefore follow Holm and rewrite the model as $Y_i = \alpha + \Sigma_{j=1}^{p}(x_{ij} - \bar{x}_{.j})\beta_j + \varepsilon_i$ or in vector form

$$\mathbf{Y} = \alpha + \mathbf{X}\beta + \varepsilon,$$

where $\mathbf{X}$ has no constant column, and mean zero in all columns, $\beta = (\beta_1, \ldots, \beta_p)'$. We will also restrict our analysis to functions $g(\beta)$ of $\beta_1, \ldots, \beta_p$ only.

Also for the so-restricted formulation we have the least squares estimate $\hat{\beta} = (\mathbf{X}'\mathbf{X})^{-1}\mathbf{X}'\mathbf{Y}$ and the above bootstrap formulas apply.

For every generated $\hat{\beta}^* = \hat{\beta}^*(\beta)$ there is a value β_c of the vector β where $\hat{\beta}^*(\beta_c) = \hat{\beta}$. Usually this solution is unique now that we have got rid of the constant. For small variations around this β_c any non-degenerate function $g(\hat{\beta}^*(\beta_c))$ will cross the value $g(\hat{\beta})$. Using (6.50) and the new version of $\mathbf{X}$, we find this critical β_c from

$$\hat{\beta} = \beta_c + (\mathbf{X}'\mathbf{X})^{-1}\mathbf{X}'\mathbf{Y}^\diamond - (\mathbf{X}'\mathbf{X})^{-1}\mathbf{X}'\mathbf{X}^\diamond\beta_c,$$
$$(\mathbf{X}'\mathbf{X} - \mathbf{X}'\mathbf{X}^\diamond)\beta_c = \mathbf{X}'\mathbf{X}\hat{\beta} - \mathbf{X}'\mathbf{Y}^\diamond = \mathbf{X}'\mathbf{Y} - \mathbf{X}'\mathbf{Y}^\diamond.$$
$$\beta_c = (\mathbf{X}'\mathbf{X} - \mathbf{X}'\mathbf{X}^\diamond)^{-1}(\mathbf{X}'\mathbf{Y} - \mathbf{X}'\mathbf{Y}^\diamond), \qquad (6.51)$$

where the inverse is supposed to exist. There is always a small probability of singularity, for example if the resampling gives the original data in the same order.

Let $g_{(1)} \leq g_{(2)} \leq \cdots \leq g_{(N)}$ be the ordered values of $g(\beta_c)$ in N bootstrap simulations. In the scale given by $g(\beta)$ we will have $g(\hat{\beta})$ in the central part (probability $c = 1 - 2\alpha$) of the distribution for $g(\hat{\beta}^*)$ for all β such that

$$g_{(N\alpha)} \leq g(\beta) \leq g_{N(1-\alpha)}$$

where $N\alpha$ is rounded to the nearest integer. This is the confidence interval for $g(\beta)$. One-sided intervals with confidence level $1 - \alpha$ are of course given from each single inequality.

Using the approach and assumptions of Freedman (1981), Holm (1990) proves that $\sqrt{n}$ times the right member of (6.51) has the same asymptotic normal distribution as the corresponding expression with the real data. This gives the asymptotic validity of the procedure. Holm also gives some examples with intervals for the β-coefficient in simple lin-

Table 6.1 *Original data (generated by simulation) and first resample of residual generating* $\mathbf{X}^\circ, \mathbf{Y}^\circ$

X		Y	X$^\circ$		Y$^\circ$
−9.5	−142.5	5.4	−0.5	−43.5	12.0
−8.5	−139.5	11.3	−7.5	−134.5	15.8
−7.5	−134.5	15.8	−4.5	−107.5	4.7
−6.5	−127.5	6.6	7.5	180.5	23.3
−5.5	−118.5	3.7	2.5	25.5	21.4
−4.5	−107.5	4.7	7.5	180.5	23.3
−3.5	−94.5	5.6	−3.5	−94.5	5.6
−2.5	−79.5	11.6	−8.5	−139.5	11.3
−1.5	−62.5	11.9	6.5	145.5	18.0
−0.5	−43.5	12.0	1.5	0.5	12.4
0.5	−22.5	7.3	−4.5	−107.5	4.7
1.5	0.5	12.4	−3.5	−94.5	5.6
2.5	25.5	21.4	−6.5	−127.5	6.6
3.5	52.5	17.7	7.5	180.5	23.3
4.5	81.5	17.1	7.5	180.5	23.3
5.5	112.5	19.6	−5.5	−118.5	3.7
6.5	145.5	18.0	−9.5	−142.5	5.4
7.5	180.5	23.3	4.5	81.5	17.1
8.5	217.5	26.0	−6.5	−127.5	6.6
9.5	256.5	33.1	5.5	112.5	19.6

ear regression which demonstrate very good results for this method compared to direct residual resampling.

Example 6.6
In the model $Y = \beta_0 + \beta_1 t + \beta_2 t^2 + \varepsilon$ we need an interval for $-\beta_1/2\beta_2 = g(\beta)$. We generate data for $\beta_0 = 10$, $\beta_1 = -1$, $\beta_2 = 0.1$, $t_i = 1, \ldots, 20$, ε_i independent with $f(x) = \exp(-(x+4)/4)/4$, $x > -4$, rounded to one decimal. The true value is $g(\beta) = 5$.

For the original data we estimate $\hat{\beta} = (-0.88, 0.09)'$, $\hat{g} = 4.67$. The resampled estimate is never seen, our new 'data' only show the same random draw as it is made for the true

ε-errors, and are needed in the computations. The value of β which makes $\hat{\beta}^*(\beta) = \hat{\beta}$ is $\beta_c = (0.0364, 0.0602)'$, and $g_0 = -0.3022$. We repeat this analysis 1000 times and get the 90% interval $-0.62 < g < 6.87$. A rather huge simulation draws the original data 10000 times independently, and makes the bootstrap confidence interval as above. The observed error frequencies were 5.75% and 2.13% at the lower and higher limits.

6.7.6 Varying variance

When the errors ε_i are not equally distributed, and for example have very different variances, we cannot neglect this in our resampling. This situation is called heteroscedasticity. Vector resampling, as discussed in Section 6.7.2, can be used. Here we will indicate a residual approach.

Let $Y_i = \mu_i(\mathbf{x}_i; \beta) + \varepsilon_i$. Estimate β and find $e_i = Y_i - \mu_i(\mathbf{x}_i; \hat{\beta})$, $i = 1, \ldots, n$. If we have no good model for the variance, a very crude resampling method is to let every $\varepsilon_i^* = \pm e_i$ with probability $1/2$ each. For the linear model, Wu (1986) has suggested the following improved idea. Let $Y_i = \mathbf{x}_i'\beta + \varepsilon_i$, and let e_i be the least squares residual for the same data. Resample Y_i as

$$Y_i^* = \mathbf{x}_i'\beta + \frac{e_i}{\sqrt{1 - w_i}} t_i^*$$

where $w_i = \mathbf{x}_i'(\mathbf{X}'\mathbf{X})^{-1}\mathbf{x}_i$, and $1 - w_i$ is the diagonal element introduced in (6.48), and t_i^* is random with $E[t_i^*] = 0$, $\mathrm{Var}(t_i^*) = 1$. Wu suggests $t_i^* = \pm 1$ together with a special (Hadamard) design for different values as one alternative, and independent sampling of t_i^* from the values $(e_j - \bar{e})/\sqrt{\Sigma(e_k - \bar{e})^2/n}$, $j = 1, \ldots, n$, as another alternative. Of course other possibilities exist on the same theme.

Resampling in such situations generally seems to require rather a lot of data for good performance, but this is typical when information about variance is important.

If observations can be grouped with approximately constant variance in each group it is natural to resample the residuals within each such group. An even better situation is the case where we have multiple observations in the design points. Then resampling of Y-values within each design point can be convenient and well motivated. The main advantage here with multiple observations is that no model assumptions are needed for this resampling.

6.7.7 Resampling in generalized linear models

Suppose we have a parametric model with one or more conventional parameters expressed as functions of a linear regression $\mathbf{x}'\beta$ in the predictors $\mathbf{x}$. In this situation parametric resampling is a natural candidate. We simply estimate β, and draw new data from the distributions defined by this parameter. This works for the class of generalized linear models, and also for some models outside the class. However, if the model is crude, such resampling can be misleading, so we warn against unqualified use of this idea.

Suppose for example that we have data $(y_i, \mathbf{x}_i)$, $1 \leq i \leq n$, and a Poisson model such that

$$Y_i \text{ is } Po(\lambda_i), \quad \lambda_i = e^{\mathbf{x}'_i \beta}.$$

We estimate β, typically by maximum likelihood, and draw a resample $Y_1^*, \ldots, Y_n^*$ for the same design points $\mathbf{x}_1, \ldots, \mathbf{x}_n$. Here Y_i^* is $Po(\hat{\lambda}_i)$ with $\hat{\lambda}_i = \exp(\mathbf{x}'_i \hat{\beta})$. The same idea can of course be used for all parametric models of this kind.

As an example where this method is dangerous, Draper and Smith (1981, p. 191) have a data set giving the number of traffic deaths in the different states of the US during one year together with a set of predictors. The predictors can only explain part of the variation between states, and the remaining variation is far too big to be explained by random variation in the Poisson distribution itself. Parametric resampling with the fitted $\hat{\lambda}_i$ will not be representative for the random variation in casualties for example from one year to another. However, the number of such accidents ought to be close to Poisson distributed in each state, and the number of casualties is high in almost every state. A more valid resample in this case is therefore from Poisson distributions with the observed values as parameters.

Another approach is to define residual resampling in generalized linear models. In order to describe one of the versions we need some basic facts about the model class. Generalized linear models are based on classical distributions like the Poisson, binomial, exponential, gamma, normal and a few more. The theory is well described in McCullagh and Nelder (1989), and an instructive introduction is also given by Dobson (1990). All the distributions considered belong to a family of distributions which can be parametrized in the form

$$f(y; \theta, \phi) = \exp((y\theta - a(\theta) + b(y, \phi))/\phi), \qquad (6.52)$$

although they often have another standard parametrization. When ϕ is known, this is the one parameter exponential family of distributions. The function f can be read both as a density and as a discrete probability function depending on the nature of the variable.

When applied to a set of data, where each data value y_i is related to a set of predictors $\mathbf{x}_i$, we introduce a linear expression $\eta_i = \mathbf{x}_i'\beta$ and a monotone link function g such that $g(\mu_i) = \eta_i$ where μ_i is the model's expected value of y_i. We let ϕ be constant for all data. Then there must be a function relating θ and η.

Some useful properties for this family are seen from the following computations, which can be made both for discrete and continuous variables but are written as integrals here. We have the total probability

$$\int f(y; \theta, \phi)dy = \int e^{(y\theta - a(\theta) + b(y, \phi))/\phi}dy = 1.$$

Differentiating this on θ we get

$$\int \frac{y - a'(\theta)}{\phi} f(y; \theta, \phi)dy = 0,$$

and solve

$$\mu = \int y f(y; \theta, \phi) dy = a'(\theta). \qquad (6.53)$$

This shows both that μ and θ are functionally related and also that ϕ is not related to the parameter θ or to μ, η, β. Instead ϕ is a variance factor as another differentiation will show,

$$\frac{d^2}{d\theta^2} \int f(y; \theta, \phi) dy = \int \left(\frac{-a''(\theta)}{\phi} + \frac{(y - a'(\theta))^2}{\phi^2} \right) f(y; \theta, \phi) dy = 0.$$

We solve the variance

$$\mathrm{Var}(Y) = \int (y - a'(\theta))^2 f(y; \theta, \phi) dy = \phi a''(\theta). \qquad (6.54)$$

Having seen the functional relations between μ, η and θ, it is time to estimate β by maximum likelihood. The steps in this solution will suggest one definition of residuals and a corresponding resample. The log likelihood for the data is

$$l = \sum \frac{y_i \theta_i - a(\theta_i) + b(y_i, \phi)}{\phi}$$

where $\theta_i = \theta_i(\eta_i)$ and $\eta_i = \mathbf{x}_i' \beta$. The maximum likelihood estimate of β can now be found by a Newton-Raphson iterative solution. The vector of derivatives with respect to $\beta_1, \beta_2, \ldots$ can be written

$$\frac{\partial l}{\partial \beta} = \sum \frac{\partial l}{\partial \theta_i} \frac{\partial \theta_i}{\partial \eta_i} \frac{\partial \eta_i}{\partial \beta} = \sum \frac{y_i - a'(\theta_i)}{\phi} \frac{\partial \theta_i}{\partial \eta_i} \mathbf{x}_i.$$

The equation $\phi \frac{\partial l}{\partial \beta} = 0$ gives, in matrix form,

$$\mathbf{X}'\mathbf{D}(\mathbf{y} - \mu) = 0,$$

where $\mathbf{D} = \mathrm{diag}(\frac{\partial \theta_i}{\partial \eta_i})$, $\mathbf{y} = (y_1, \ldots, y_n)'$, $\mathbf{X}' = (\mathbf{x}_1, \ldots, \mathbf{x}_n)$. (The symbol diag will be used with several meanings. If the argument is a vector we take the matrix with this diagonal and zeros elsewhere, if the argument is a square matrix we take either the vector of diagonal elements or the matrix with the same diagonal but zeros elsewhere depending on which version we need for the formulas.) Differentiating once more, we have in matrix form

$$\phi\frac{\partial^2 l}{\partial\beta\partial\beta'} = \sum\left((y_i - a'(\theta_i))\frac{\partial^2\theta_i}{\partial\eta_i^2} - a''(\theta_i)(\frac{\partial\theta_i}{\partial\eta_i})^2\right)\mathbf{x}_i\mathbf{x}_i'$$

with the expected value

$$\phi E[\frac{\partial^2 l}{\partial\beta\partial\beta'}] = -\mathbf{X}'\mathbf{DADX}, \qquad (6.55)$$

where A is diagonal with element $a''(\theta_i)$ at position (i, i).

Given some starting vector β^0, Newton-Raphson's iterative solution gives recursively

$$\beta^{t+1} = \beta^t + (\mathbf{X}'\mathbf{DADX})^{-1}\mathbf{X}'\mathbf{D}(\mathbf{y} - \hat{\mu}) \qquad (6.56)$$

with $\mathbf{D}$ and $\mathbf{A}$ computed at the parameter $\hat{\beta}^t$ in each step. (If this is heavy, we can sometimes use the inverse more than once.) Let $\hat{\beta}$ be the point of convergence.

The $\mathbf{y} - \hat{\mu}$ can be seen as residuals, but the corresponding variables $\mathbf{Y} - \hat{\mu}$ do not have equal variance. We know from (6.54) that $\phi\mathbf{A}$ has the variances of $\mathbf{Y}$ in the diagonal. Define the Pearson residuals as $\mathbf{A}^{-\frac{1}{2}}(\mathbf{y} - \hat{\mu})$ with elements $(y_i - \hat{\mu}_i)/\sqrt{a_i''(\theta_i)}$. The variances are now closer but not equal since we have not considered the uncertainty of $\hat{\mu}$. Moulton and Zeger (1991) therefore suggest another set of residuals for the resampling and state that these always outperformed the Pearson residuals in their simulations. We can explain their version by computing the approximate variance of $\hat{\mu}$. Since $\mu_i = a'(\theta_i) = a'(\theta_i(\eta_i(\beta)))$, a linear approximation gives

$$\text{Var}(\hat{\mu}_i) = (a''(\theta_i))^2 \left(\frac{\partial \theta_i}{\partial \eta_i}\right)^2 \left(\frac{\partial \eta_i}{\partial \beta}\right)' \mathbf{C}_{\hat{\beta}} \frac{\partial \eta_i}{\partial \beta},$$

with all derivatives computed at the parameter point $\hat{\beta}$ and with

$$\mathbf{C}_{\hat{\beta}} = -\left(\frac{\partial^2 \ln l}{\partial \beta \partial \beta'}\right)^{-1} = \phi (\mathbf{X}'\mathbf{D}\mathbf{A}\mathbf{D}\mathbf{X})^{-1}$$

according to standard asymptotic theory, see Cox and Hinkley (1974, Chapter 9). For the full vector this gives

$$\text{Var}(\hat{\mu}) = \phi \mathbf{A} (\mathbf{A}^{\frac{1}{2}}\mathbf{D}\mathbf{X}(\mathbf{X}'\mathbf{D}\mathbf{A}\mathbf{D}\mathbf{X})^{-1}\mathbf{X}'\mathbf{D}\mathbf{A}^{\frac{1}{2}}).$$

Let $\mathbf{W} = \text{diag}(\mathbf{A}^{\frac{1}{2}}\mathbf{D}\mathbf{X}(\mathbf{X}'\mathbf{D}\mathbf{A}\mathbf{D}\mathbf{X})^{-1}\mathbf{X}'\mathbf{D}\mathbf{A}^{\frac{1}{2}})$. By analogy with linear regression the two variances *subtract* in the residual, and we therefore divide each residual by the square root of the corresponding diagonal element of $\mathbf{A} - \mathbf{A}\mathbf{W}$ to make the variances equal. Let

$$\mathbf{E} = (\mathbf{A} - \mathbf{A}\mathbf{W})^{-\frac{1}{2}}(\mathbf{y} - \hat{\mu}). \tag{6.57}$$

Draw a resampled vector $\mathbf{E}^*$ from the elements of $\mathbf{E}$. Let $(\mathbf{Y} - \hat{\mu})^* = \mathbf{A}^{\frac{1}{2}}\mathbf{E}^*$ to make the variance equal to $\phi \mathbf{A}$, the variance of $\mathbf{Y}$ (and not of $\mathbf{Y} - \hat{\mu}$). The bootstrap estimate $\hat{\beta}^*$ is defined by a single Newton-Raphson step in Moulton and Zeger (1991),

$$\hat{\beta}^* = \hat{\beta} + (\mathbf{X}'\mathbf{D}A\mathbf{D}\mathbf{X})^{-1}\mathbf{X}'\mathbf{D}\mathbf{A}^{\frac{1}{2}}\mathbf{E}^*, \tag{6.58}$$

with all matrices computed at the parameter $\hat{\beta}$.

We have avoided defining $\mathbf{Y}^*$, since $\hat{\mu} + \mathbf{A}^{\frac{1}{2}}\mathbf{E}^*$ can be impossible values for $\mathbf{Y}$ when we have a discrete distribution.

There are some other definitions of residuals as discussed in Pierce and Schafer (1986) or Hinkley, Reid and Snell (1991, Chapter 4), but the resampling is most direct for residuals of the type we have discussed here.

6.7.8 Model selection and resampling

In model selection problems we have the added complication that no model is generally accepted. Some of the resampling methods will therefore fail in this situation. A method which always works is vector resampling, but it has the drawbacks already mentioned that the variations of the design matrix will enter into the measures. However, with large data sets without very influential observations the method is quite useful.

The classical statistical measures of uncertainty are computed for the selected model only. That is exactly what we want to avoid when computer intensive methods are applied to model selection problems. Thus comparing residual resampling from different models is not a possible way to proceed. However, if all the models can be seen as special cases of one estimable maximal model, and this model is regarded as 'true', then residual resampling from the maximal model will provide a valid basis for conclusions about the entire model selection and estimation process. There is no need for the maximal model to be one of the candidate models in the selection.

Model selection was illustrated in Chapter 3 on the cement hardening data of Table 3.1. Formulas (3.10) and (3.11) defined the performance measure $CMV(p)$ as a function of the model size, and p_0 was the model size minimizing CMV. Such results are of course uncertain and can be studied by bootstrap methodology. We will here apply parametric bootstrap to this problem, under the assumption that the full model $\mathbf{Y} = \mathbf{X}\beta + \varepsilon$ is true with independent $N(0, \sigma)$-distributed components ε_i in ε. With only minor modifications we can use standardized residuals or projected residuals instead. (Non-standardized residuals are too inflated since we have so many parameters compared to the number of data values, abstract residuals give confidence intervals but will not show all interesting aspects here, at least not without some extra development of the theory, and vector resampling gives too much design variation in our small data set.)

Table 6.2 *Bootstrap selected model structures on cement data in 1000 replicates. Variable 1, constant, is forced into all models.*

Non-constant predictors	Model size	Frequency
2 3	3	49
2 5	3	198
2 3 4	4	108
2 3 5	4	358
2 4 5	4	142
2 3 4 5	5	145

Resampling 1000 times and performing the full model selection analysis as forward selection for each resample takes less than a minute on a Sun Sparc computer for this small data set. It is therefore no problem to apply the analysis to reasonably large data sets. (Time increases linearly with the number of data values, less than quadratically with the number of variables, but more close to cubic with the maximal model size studied.) The bootstrap simulations show that the model size decision varies randomly, and within the model sizes 3 and 4 (with 2 and 3 non-constant predictors) also some different model structures are possible and will be selected. All selected model structures and their frequencies are displayed in Table 6.2.

Other interesting results are read from the bootstrapped $CMV(p)$ functions. Of the 1000 resamples, 200 are displayed in Figure 6.4 and are drawn interpolated between integer model sizes.

The general impression for these data is that the model selection has substantial randomness and that the model size decision can be dependent on the random errors. The CMV-curves are very flat for $3 \leq p \leq 5$ and give no hope for significant differences there. Another more important impression is that the bootstrap combines favourably with the validation methods with useful uncertainty measures as a result.

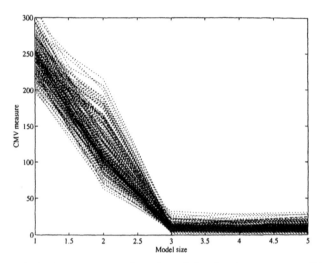

Figure 6.4 *Bootstrap results of model selection measure* CMV *(p) on cement data.*

Any alternative automatic modelling method can of course be analysed in the same way, and there are a few different suggested methods in the regression literature. Also quite different methods can be compared and tested for significant differences in performance.

We are now in a position where for example the F-tests in stepwise regression can be replaced by more valid bootstrap measures of significant improvement when another parameter is selected in the model. The bootstrap analysis in Figure 6.4 indicates for example that the modelling improves significantly when p increases from 1 to 2 and from 2 to 3. The improvement is by no means clear when p increases from 3 to 5. For this data set the situation is so clear cut that we hardly need a more careful analysis than just looking at the plot.

We might in other cases attempt to sharpen the analysis with some studentized interval for

$$(\mathrm{CMV}^*(p_1) - \mathrm{CMV}^*(p_2))/\hat{\sigma}^*(\mathrm{CMV}(p_1) - \mathrm{CMV}(p_2)),$$

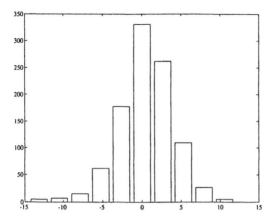

Figure 6.5 *Resampled differences* CMV *(3)*–CMV *(4)*– $\mathbf{d}_3^{*\prime}\mathbf{d}_3^{*}/n +$ $\mathbf{d}_4^{*\prime}\mathbf{d}_4^{*}/n$ *for the cement data.*

but have difficulties in producing $\hat{\sigma}^*(.)$. The simple and the percentile intervals are probably the maximal effort we can make in this situation. Another difficulty here is to define what we are making an interval for. There is no clear parameter $CMV(p_1) - CMV(p_2)$, since the expected value is of limited interest here. We are rather in a prediction situation where we want conclusions about the outcome of this difference for the original data with the aid of a bootstrap simulation. For each resampled data set we can compute the mean square prediction errors of models p_1 and p_2 in the forward selection (that is how these models would perform if they were fixed and had to predict new resampled data). This becomes a pure mean square bias difference $\mathbf{d}_{p_1}^{*\prime}\mathbf{d}_{p_1}^{*}/n + s^2 - \mathbf{d}_{p_2}^{*\prime}\mathbf{d}_{p_2}^{*}/n - s^2$ where $\mathbf{d}_p^* = \mathbf{X}_{[p]}^*\beta_{[p]}^* - \mathbf{X}\hat{\beta}$ is the bias of model p_1 in the bootstrap simulations and the index $[p]$ denotes the structure selected for model size p. We can also compute $CMV^*(p_1) - CMV^*(p_2)$ for each such case (the differences are contained in Figure 6.4), and study the differences between starred CMV differences and starred mean square bias differences, and this version can be handled exactly as ordinary bootstrap differences. See Figure 6.5. Although the old cement data are of little practical concern today, the modern computer intensive methods show their

force and ability to handle complex analyses on both this
old problem and the many new problems.

6.8 Bootstrap realizations of a stationary process

The bootstrap method, started as a technique for independent and identically distributed data, was later formulated
for regression problems as discussed earlier, and has also been
extended to certain time series problems. The first approach
in these extensions has been to compute the time series residuals, as one step ahead prediction errors from some model,
and resample among these. Other possibilities are to resample in the spectral domain, or to subdivide data into approximately independent blocks as in Künsch (1989). (This takes
much data.) In the rare case where several independent realizations are observed, these can be regarded as independent
vector observations and resampled accordingly.

An important borderline is between situations with well
defined parametric models, and situations where a model is
searched, or the analysis has to be based on very general
properties like stationarity.

Given that bootstrap replicates can be constructed in a
sensible way from a given observed realization, what can we
use them for? This question seems to be particularly relevant
in the case of stochastic processes since there is sometimes
an interest in trying to extend a too small sequence of data
and one could believe that bootstrapping provides a solution
to this problem. Let us first recall the role of bootstrap analysis in general. The resampled time series does not give any
new information to the observer since only true observations
have information on the true system. Instead the bootstrap
will extract the information of the original data. The role
of the bootstrap is therefore best seen as a very implicit definer of functions which are natural and useful as estimators
of parameters and uncertainties. The same view was expressed in Chapter 4. In stochastic processes most classical
problems like estimation of autocorrelations, spectral densities, ARMA parameters, level crossing probabilities, etc.
can be approached by this technique. Since the information

is given in terms of realizations, it can be directly used in the simulation of more complex systems. In problems where good analytical solutions are available, these can be compared with the bootstrap results in order to get some feeling of how accurate bootstrap results are.

6.8.1 Residual resampling

Residual resampling of a time series parallels very much the residual resampling in regression. We must structure the model as a one step ahead prediction formula and a random noise term. This can be made in many ways, but usually rather simple autoregressive structures or more general ARIMA models will be used. However, the most interesting future development is probably outside this class, perhaps in the direction of multivariate and non-linear situations. Suppose we have an autoregressive model

$$Y_t = \sum_{i=1}^{p} a_i Y_{t-i} + \varepsilon_t$$

with ε_t independent and identically distributed as usual. Let $y_1, \ldots, y_n$ be a sequence of data, considered as coming from the model. We can then estimate a_i by least squares for example, and compute $e_t = y_t - \Sigma_1^p \hat{a}_i y_{t-i}$ for $t = p+1, \ldots, n$. A resample ε_t^* can be drawn as independent variables from the empirical distribution of $e_{p+1}, \ldots, e_n$, and a resample of the process is generated recursively by the relation $Y_t^* = \Sigma_1^p \hat{a}_i Y_{t-i}^* + \varepsilon_t^*$. The resample is not restricted to $t \leq n$, but can go on unboundedly, but for the evaluation of inferences made from the sample, the resample will usually have the corresponding length. Starting values $Y_1^*, \ldots, Y_p^*$ are necessary and their construction should mimic the situation for the original data. The most common case is when the model is stationary and we start with the stationary distribution. Then we can start anywhere (perhaps with p zeros), and run the estimated bootstrap model to approximate stationarity before we define time to be zero and continue from that time on. The same approach works for any model which can be transformed to $Y_t = g_t(Y_{t-1}, Y_{t-2}, \ldots; \theta) + \varepsilon_t$, where ε_t are equally distributed and independent of the past observations, where g_t is known, and θ can be estimated. The bootstrap procedure in these models is simply a simulation of the estimated model with residuals from the empirical distribution. In Freedman (1984) it is shown that the bootstrap gives asymptotically valid error distributions of the estimated coefficients in stationary linear econometric models. Bose (1988) improved the bootstrap approximations by Edgeworth corrections and showed improved asymptotic behaviour, Basawa et al. (1989) showed correct asymptotic behaviour for the bootstrap of least squares estimates in nonstationary autoregressive models.

When the model is unknown, but belongs to a class of parametric models, we cannot use the approach above directly. We have exactly the same problems here as with regression models. It is for example not fair to resample each model by itself and compare the so-generated performance. Suppose we have to select one of the models on the basis of the data, and then estimate or predict by that model. If a useful model exists such that all other models are special cases, then one possibility is to resample in the large model and study the combined model selection and estimation procedure. This works for instance for the class of autoregressive models of order at most $p_{\max}$, $Y_t = a_1 Y_{t-1} + \ldots + a_p Y_{t-p} + \varepsilon_t$, $1 \leq p \leq p_{\max}$. However, it does not always work in practice for the popular class of ARMA models (or ARIMA) since high order models do not always give useful estimates in the stationary and invertible range by standard methods. By careful definition of the estimation and selection procedure the problems can be handled for important parameters with a global (over all models) meaning.

6.8.2 Spectral resampling

For stationary processes, the spectral representation provides an alternative. The advantage is that time dependency is transformed into uncorrelated spectral increments, and at least for Gaussian processes this means independence. Another advantage is that this is a non-parametric method for the entire class of (weakly) stationary processes.

The idea of resampling in the spectral domain has appeared fairly recently. For the special case where the spectral density itself is estimated one such approach was analysed by Franke and Härdle (1992). A more general construction of bootstrap realizations was suggested in Nordgaard (1990, 1992) and opens a wide class of new applications to this kind of bootstrap analysis. According to Cressie (1991) the same idea was developed independently by Hurvich and Zeger (1987).

Let x_t, $-n \le t \le n$, be an observed sequence from a stationary, real valued stochastic process in discrete time with covariance function $R(\tau)$ allowing a continuous spectral density $S(\omega)$ and with the expected value zero. We have for simplicity assumed an odd number of time points. Define the natural frequencies $\omega_k = k\frac{2\pi}{2n+1}$, $-n \le k \le n$, for a series of this length. Let

$$b_k = \frac{1}{2n+1} \sum_{t=-n}^{n} x_t \cos(\omega_k t)$$

$$c_k = \frac{1}{2n+1} \sum_{t=-n}^{n} x_t \sin(\omega_k t)$$

and

$$a_k = b_k - i\,c_k = \frac{1}{2n+1} \sum_{t=-n}^{n} x_t e^{-i\omega_k t}. \qquad (6.59)$$

Now there is a spectral representation of the observed sequence so that

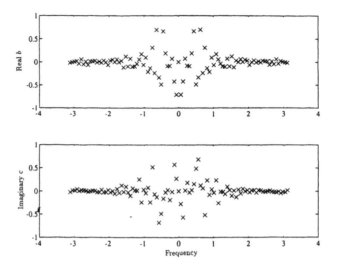

Figure 6.6 *Real part b_k above, and imaginary part $-c_k$ of spectral components for transformed sun spot numbers.*

$$x_t = \sum_{k=-n}^{n} a_k e^{i\omega_k t} \qquad (6.60)$$

for $-n \leq t \leq n$. It can easily be proved that b_k and c_k are uncorrelated. Also the a_k are weakly correlated and asymptotically uncorrelated for separated frequencies. However, since ω_k and ω_{-k} are the same frequency we have $a_{-k} = \overline{a_k}$. For a Gaussian process b_k and c_k are independent with approximately the same variance and mean zero. This is usually a good approximation also for non-Gaussian processes. The squared amplitude $|a_k|^2 = b_k^2 + c_k^2$ has therefore an exponential distribution (the chi-squared distribution with 2 degrees of freedom is exponential with mean 2) which is used by Franke and Härdle (1992) for the resampling of spectral density estimates.

In Figure 6.6 we show the real and imaginary parts of a_k for a sequence of transformed sunspot numbers (1771–1869). In Figure 6.7 we show the same thing for one week ahead prediction errors of stock prices for the Swedish company SCA

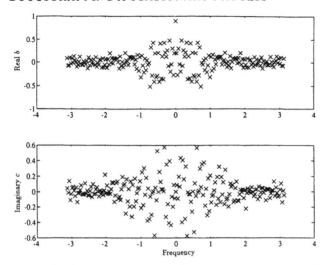

Figure 6.7 *Real part b_k above, and imaginary part $-c_k$ of spectral components for errors of one week ahead stock predictions.*

during 1985. The predictions are described in Chapter 7. The general impression of such plots is their complete randomness and a varying variance. Except perhaps for a_0, the components have mean zero and a variance proportional to the spectral density at the frequencies. (If the process is sampled from a process in continuous time, the spectral density of the sampled process can be an alias mixture of the spectrum for the continuous time process). Only the second process has the expected value close to zero (since no sensible prediction method with square error loss will allow a non-zero average in the long run). For the sun spot data we have forced a_0 to be zero by subtracting the mean value from the series.

If we can find a good way to resample the a_k, a resampled realization is achieved from the a_k^* and formula (6.60) as

$$X_t^* = \sum_{k=-n}^{n} a_k^* e^{i\omega_k t}, \quad -n \le t \le n. \qquad (6.61)$$

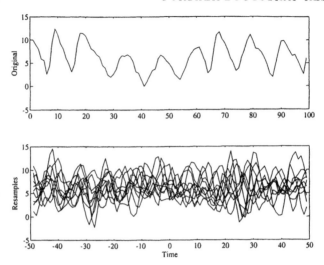

Figure 6.8 *Square root of sunspot numbers from 1770–1868, above. Five resampled realizations with a_0^* generated as $N(0, \hat{\sigma}_0)$ and mean of original data added, below.*

From this realization all estimates and other computations made on the original data will be bootstrapped by the same computations on X_t^*.

Nordgaard (1990, 1992) studies some different versions of resampling the a_k and checks their stationarity together with the convergence of process covariances to the correct limit. In the simplest version a_k^* is drawn at random and independently for different $k > 0$ from the elements of the matrix $A_k = \alpha_k (1 - 1 \quad i - i)$. Here

$$\alpha_k = \begin{cases} (a_{k-1} \ a_k \ a_{k+1})' & \text{if } 0 < k < n; \\ (a_{n-1} \ a_n \ a_n)' & \text{if } k = n; \end{cases} \tag{6.62}$$

and $a_{-k}^* = \overline{a_k^*}$.

The value a_0^* is special and must be real. If we are sure that $E[X_t] = 0$, we can resample a_0^* from the values $\pm a_0, \pm |a_1|$ perhaps by unequal weights as long as the expected value remains zero. If the process mean can be different from zero, both a_0^* and a_1^* have to be treated differently since a_0 is the expected value plus random variation, and we only want the random part in our resampling. Instead use for example $\alpha_1 =$

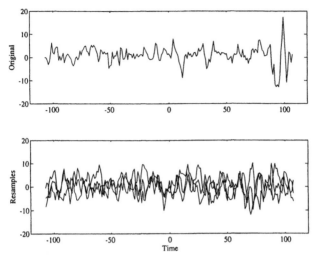

Figure 6.9 *Five days ahead prediction errors of stock values for the company SCA during 1985, above. Three resampled realizations by the EC method, below.*

$(a_1\ a_2)'$ only, and compute an estimate $\hat{\sigma}_0$ for a_0 from $\sigma_0^2 = \text{Var}(a_0) = \text{Var}(\bar{X})$, with $\text{Var}(\bar{X}) = \frac{1}{2n+1}\Sigma_{\tau=-n}^{n}R(\tau)(1 - \frac{|\tau|}{2n+1})$, where we put in an estimate $\hat{R}(\tau) = \frac{1}{2n+1}\Sigma(x_{t+\tau} - \bar{x})(x_t - \bar{x})$ summed over the useful part of the series. Let a_0^* be $N(0, \hat{\sigma}_0)$ and $X_t^* = \bar{x} + \Sigma_{k=-n}^{n} a_k^* \exp(i\omega_k t)$. In both cases the realizations X_t^* are real, since the side conditions on a_k^* will cancel the imaginary parts.

By the method (6.61) we have for $k \neq 0$ $E_* a_k^* = 0$, $E_* a_k^{*2} = 0$, $E_*[|a_k^*|^2] = (|a_{k-1}|^2 + |a_k|^2 + |a_{k+1}|^2)/3$. The resampled process has therefore $E[X_t^*] = 0$ and covariances given by

$$R^*(\tau) = E_*[X_{t+\tau}^* \overline{X_t^*}] = \sum_k \sum_r E_* a_k^* \overline{a_r^*} e^{i\omega_k(t+\tau)} e^{-i\omega_r t}$$

$$= \sum_k E_*[a_k^* \overline{a_k^*} e^{i\omega_k \tau} + a_k^* \overline{a_{-k}^*} e^{i2\omega_k t + i\omega_k \tau}]$$

$$= \sum_{k=-n}^{n} E_*[|a_k^*|^2] e^{i\omega_k \tau}$$

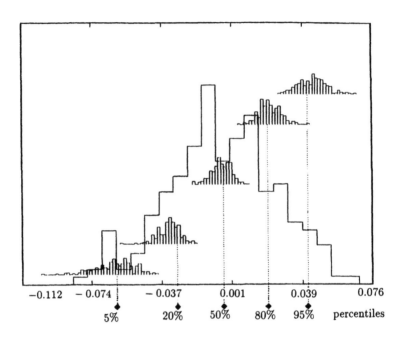

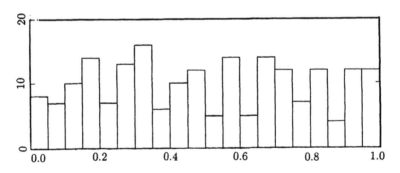

Figure 6.10 *Histogram of simulated* (X_{t+1}, X_t)-*correlation esti-mates, with small histograms for the resampled estimates of 5, 20, 50, 80, 95% percentiles, above. Histogram of the original esti-mate's position in the resampling distribution, below. 200 samples of length 501 from* $X_t = 0.7X_{t-1} - 0.2X_{t-2} + \varepsilon_t$ *are each resampled 100 times. From Nordgaard (1990).*

since $a_k^* \overline{a_{-k}^*} = a_k^{*2}$ has the expected value zero. (Taking the complex conjugate of the real X_t^* changes nothing but gives a better structure to the internal complex computations, and it is the standard definition for complex processes.) This result shows covariance stationarity and parallels a corresponding development for the theoretical $R(\tau)$.

Another resampling version, which is superior in several simulated cases, is to resample values $|a|_k^*$ from the absolute values of the elements of α_k in (6.62), and then draw independent $\mathrm{Re}(-\pi, \pi)$-distributed variables ϕ_{1k}, ϕ_{2k} and let

$$a_k^* = |a|_k^*(\cos \phi_{1k} + i \sin \phi_{2k}), \quad k = 1, \dots, n, \qquad (6.63)$$

and as usual set $a_{-k}^* = \overline{a_k^*}$.

This method gives both random phase and a continuous amplitude variation, but preserves the expected effect ($=$ variance) contribution from the frequency ω_k. This resampling method is called EC (extended circle) by Nordgaard, to distinguish it from from another C method which has $\phi_{2k} = \phi_{1k}$ and no smooth amplitude variation. The EC-method gives excellent confidence intervals for correlations and spectral estimates for signals of 501 data values from the model $X_t = 0.7X_{t-1} - 0.2X_{t-2} + \varepsilon_t$. See Figure 6.10.

6.9 Exercises

Exercise 6.1   Make a parametric bootstrap with the class of normal densities and the parameters (a) μ, (b) σ, (c) $\mu + 2\sigma$, (d) μ/σ. Where possible, compare the conclusions from large bootstrap simulations with the conclusions from traditional analysis.

Exercise 6.2   Make a bootstrap analysis of the estimate $\hat{\lambda}$ maximizing the Chi-square function $T(\lambda)$ in Example 6.2.

Exercise 6.3   In Example 6.3 the variables Y and Z are integers such that $0 \leq Z \leq Y \leq 100$. It is possible to replace all simulations by exact computations, but how should it be organized? Is the amount of computation comparable to that of the simulations in the example?

Exercise 6.4   From a sample of n independent $N(\mu, \sigma)$-distributed variables a 95% confidence interval for μ is (incorrectly) computed as $\bar{x} \pm 1.96s/\sqrt{n}$, where $s^2 = \Sigma(x_i - \bar{x})^2/(n-1)$. Check first the level of this interval, then apply some version of Loh's level adjustment method and find the true level of the new interval. Use some sample size around $10 \leq n \leq 20$ and tables of the Student and the normal distribution (simulations are unnecessary).

Exercise 6.5   A bivariate time series $(x(t), y(t))$, $1 \leq t \leq n$ is observed and seems stationary. Discuss methods to resample the process. What could parametric resampling mean in this case?

Exercise 6.6   Perform residual bootstrap and vector resampling on the data in Table 6.1 (first three columns).

CHAPTER 7

Computer intensive applications

In this chapter we apply validation and bootstrap methods to some problems in different fields. The selected problems are taken from work where we have been involved in Linköping and build on contacts with departments and institutes like the Road and Traffic Institute, the Meteorological and Hydrological Institute, the Military Weather Service, and medical departments. Hopefully this sample of studies will convey the message that these methods are very generally applicable and solve many problems which would have been otherwise unsolved or would have resulted in some inferior solution. Many other applications have been described by other authors in journals, and for example the references to work by Efron and co-authors contain many interesting cases.

7.1 Validation and bootstrap in road safety analysis

7.1.1 The single crossing

A road junction in the countryside is considered to be dangerous. During the years 1977–83 the following numbers of serious accidents occurred: 3, 0, 2, 1, 6, 2, 2. Less serious accidents are not well reported and provide too uncertain information. If traffic flow and other conditions are approximately constant, we can regard these figures as independent outcomes from a probability distribution (the Poisson distribution) with mean μ. The most natural estimate of μ will of course be the average $16/7 = 2.3$ if these data are the only information we have.

Since the number of accidents is regarded as high, the crossing is a candidate for modification in order to make it

Figure 7.1 *Damn driver – damn bike!*

safer. The authority in charge of this will of course evaluate the effects of the changes. Suppose that our crossing was rebuilt at the end of 1983, and that an average of 1.25 serious accidents per year occurred during 1984–87. Then a natural estimate of the effect of the rebuilding would be a decrease of $2.3 - 1.25 = 1.05$ accidents per year. This estimate is of course very uncertain due to random variations, but putting together the experiences from several similar rebuildings, the estimate can be made more stable. However, one smells a rat here. The uncertainty is not the only problem. This kind of estimated effect is systematically wrong and is severely biased towards overestimated effects of modifications. The lack of recognition of this bias seems to have caused some expensive mistakes over the years, but it is now a well-known phenomenon in traffic theory. We will explain the effect in the next section. The problem was analysed with computer intensive techniques by Junghard (1990).

7.1.2 Several crossings

Accident data have been collected from a large number of junctions over the years $1, 2, \ldots, r$. These data were all taken before any possible modification was made. Let Y_{kt} be a measure of the accidents in junction k and year t. This measure is not always the number of accidents, it may be any kind of cost associated with the accidents. One traffic flow measurement has also been made at each junction and is considered representative for all the r years. It is denoted (P_k, S_k) where P_k is the flow on the primary (busiest) road and S_k is the flow on the secondary road in junction k.

Anyone with information about junction k alone will estimate the expected danger (cost) by the average

$$\overline{Y}_k = \frac{1}{r}(Y_{k1} + \ldots + Y_{kr}) \qquad (7.1)$$

without any reference to other crossings or to the traffic flow.

The data base calls for a different approach. We may for example consider a regression model like

$$Y_{kt} = \beta_1 P_k + \beta_2 S_k + \beta_3 P_k S_k + \varepsilon_{kt} = \mu_k + \varepsilon_{kt} \qquad (7.2)$$

or some other function of the traffic flow. When Y is the number of accidents we may take Y_{kt} as Poisson distributed with mean μ_k above. However, since other factors than the traffic flow may have influence on the expected number of accidents, a different approach is often motivated. The estimated regression equation will give us a 'normal' level for the accidents at a junction. We get however, no guidance as to which crossings should be modified if precautions are made only for junctions which, due to their design, get more, or more serious, accidents than is 'normal' with respect to the traffic flow. We obviously need a measure of the accidents at junction k which can be compared to the expected value of the regression equation. We also need to compare this measure to similar measures later on if the junction is to be modified.

The only particular information we have besides the regression equation is the average $\overline{Y}_k$. However, $\overline{Y}_k$ may be high (or low) for two different reasons. One is the random fluctuations, and the other is the geometry of the crossing and other local conditions. The random fluctuations are strong and can easily dominate the other effects. If the worst crossings are selected for modification, several (perhaps most) of them will have extra high $\overline{Y}$-values due to the randomness. With or without modification these crossings will on average give fewer accidents or lower costs later on. This effect of the selection is the reason for the systematic bias mentioned above and it is not so easily described theoretically. One can show that a good estimate of the expected value in each junction is given by a linear combination of the local average and the regression model

$$\hat{\mu}_k = \alpha_1 \overline{Y}_k + \alpha_2(\hat{\beta}_1 P_k + \hat{\beta}_2 S_k + \hat{\beta}_3 P_k S_k). \qquad (7.3)$$

Sometimes the condition $\alpha_1 + \alpha_2 = 1$ is imposed. This estimate, with proper α_1, α_2, also works for a subset of junctions selected for their high $\overline{Y}$. Is $\hat{\mu}_k$ a biased estimator? The answer is yes if k is given a priori without reference to $\overline{Y}_k$, but for a random k selected due to its high (or low) $\overline{Y}_k$ the cross validation analysis below will make $\hat{\mu}_k$ closer to unbiased. Copas (1983) discusses the same kind of effect for his shrinkage estimators.

The α_1-weights are difficult to derive analytically (a lot of assumptions are needed) but they are easily determined by cross validation, arranged so that excluded data are as well predicted as possible. In Junghard (1990) the following approach was taken (some different versions were studied).

Estimation set ES_t: All data except from year t;

Test set TS_t: Year t in all junctions;

$t = 1, \ldots, r$. Let $-t$ denote the exclusion of year t. Let $\alpha_1 = \alpha$, $\alpha_2 = 1 - \alpha$. Then

$$\hat{\mu}_k(-t, \alpha) = \alpha \overline{Y}_k(-t) + (1 - \alpha)(\hat{\beta}_1(-t)P_k$$
$$+ \hat{\beta}_2(-t)S_k + \hat{\beta}_3(-t)P_k S_k). \qquad (7.4)$$

The sum

$$T = \sum_{t=1}^{r} \sum_{k=1}^{n} (Y_{kt} - \hat{\mu}_k(-t, \alpha))^2$$

is minimized with respect to α. Compare the estimation by Stone in Section 3.2. This gives an estimated optimal value $\hat{\alpha}$ and the cross validation estimates of μ_k are then given by (7.3). For the number of accidents Y, and a slightly different regression model, a sample of 458 junctions and seven years of data gave

$$\hat{\alpha} = 0.33, \qquad \sigma_{\hat{\alpha}} = 0.054. \qquad (7.5)$$

We then rely by 1/3 on the local observation and weight the regression by 2/3 for this sample. (The weight α will depend heavily on the average number of accidents per crossing and the variation in traffic flow between crossings. A more stable parameter is defined in Junghard (1990).)

7.1.3 Estimating the uncertainty of $\hat{\alpha}$

In (7.5) we gave a standard deviation for $\hat{\alpha}$. This value is of course an estimate, but how can it be estimated? Since the $\hat{\alpha}$ is given by a computer run it has a complex mathematical description. We cannot analyse this by classical methods, so we use our computer intensive technique once more. This time the bootstrap approach provides a solution. The 458 junctions are regarded as a sample of all such junctions. If we resample among these junctions we make a kind of vector resampling (compare Section 6.7.2), and get bootstrap data where the entire cross validation estimation of α can be repeated. The computer will never be bored or tired! Doing so, the standard deviation of $\hat{\alpha}^*$ becomes 0.054 as given in (7.5). In the schedule below we illustrate the two steps.

Resampling schedule

		time					
		1	*2*	*3*	$\cdots$	*r*	
	1	Y_{11}	Y_{12}				
	2						*Bootstrap*
crossing	*3*						*resampling*
number	*4*						*and estimation of*
							uncertainty of $\hat{\alpha}$.
							(Vertical
							direction.)
	$\vdots$						
	458						

Cross validation and
estimation of $\hat{\alpha}$
(Horizontal direction.)

Notice that the cross validation cannot be replaced or followed by bootstrap resampling in the time direction since this will introduce a dependency. With some large probability the predicted column or value will appear also in some of the other columns. Nor can we apply cross validation in the vertical direction when we estimate α. Possibly some bootstrap procedure can be invented to solve both the estimation and uncertainty analysis, but this is at present unsolved.

    Returning to our single junction we had an average of $\overline{Y} = 2.3$ accidents/year. If the regression equation says 1.0 accidents per year for this traffic flow, we get $\hat{\mu} = 0.33 \cdot 2.3 + 0.67 \cdot 1.0 = 1.43$. If the average after rebuilding is 1.25 we will now estimate an effect of $1.43 - 1.25 = 0.18$ accidents per year from the changes of the junction. This can be compared with the estimate 1.05 in the first section. The uncertainties are large compared to the size of this single estimate. Averaging over many cases, valid estimates can be found. This is however another story.

7.2 Forward validation on the stock market

Fluctuations of stock prices have been compared to stochastic processes of special types. Under idealized economic assumptions, today's quotations are the best possible estimates of expected future values. By value we mean discounted or inflation adjusted prices, but in the short run and with reasonable inflation the quotations themselves (in £, $, Skr) are approximately the same thing. Let $X(t)$ denote the value of a certain stock at time t. Then, by the economic idealization, a stochastic model should have the property

$$E[X(t+s)|X(t)] = X(t) \qquad (7.6)$$

for $s > 0$. However, there may be more information around than just the price of a single stock. Let $\mathcal{F}_t$ denote all the information available to the market up to time t*. A generalization of (7.6) can then be written as

$$E[X(t+s)|\mathcal{F}_t] = X(t). \qquad (7.7)$$

The probabilistic term for this property is that (7.6) or (7.7) define $X(t)$ as a martingale relative to $X(t)$ itself or to $\mathcal{F}_t$.

However, if the stock market acts as a system of martingales, the prediction of stock prices has a simple but rather uninteresting solution. The expected value, given the available information, is the optimal forecast in the least squares sense. In martingale models we can therefore make no better than take the last available price $X(t)$ as our forecast at time t of $X(s+t)$. These kinds of forecasts are called persistency forecasts in some applications and may be of some interest also for systems which do not necessarily behave as martingales, for instance in meteorology where they are often used

---

* Theoretically $\mathcal{F}_t$ represents all events defined by random variables observable up to time t, or rather the sigma-algebra, also called the Borel field by Chung (1974), generated by such events. If every possible observation corresponds to a random variable in the model, with time index corresponding to real time for the observation, then conditioning on $\mathcal{F}_t$ is equivalent to using all this information.

as baseline models to compare other hopefully better prediction models against.

Special models of the martingale type are symmetric random walks, in particular the Wiener process, where successive changes are modelled as independent normal variables
with mean zero.

The martingale structure of stock quotations is of course
an idealization, and reality can be more complex. Also people consistently try to outperform the persistency forecasts
by various systems. Many of these systems may seem efficient during limited periods of time. The overfitting problem
is obviously very acute when many such prediction methods
are studied on historical data. Careful validation is therefore
needed in order to distinguish true information from spuri-
ous.

We will demonstrate the forward validation technique on
a one year record of four stock prices accompanied by information about two indexes and two rates. There is more
useful information registered, like the volumes of sold stocks,
but for this illustration we use a more limited data base.
The aim is to illustrate methods, not 'money-making'. Let
$t = 1, \ldots, 249$ be an enumeration of the (Swedish) labour
days of 1985. Define $X_i(t)$, $i = 1, \ldots, 8$, as the following
series:

$i = 1$	STORA	stock prices
$i = 2$	SCA	stock prices
$i = 3$	MODO	stock prices
$i = 4$	HOLMENS	stock prices
$i = 5$	R30	30 days loan interest
$i = 6$	SGI	Swedish general index
$i = 7$	SFI	Swedish forest industry index
$i = 8$	GM	German marks, exchange rate.

All the four companies have their main activities in the
forestry and paper industry. Since 1985 the structure of these
companies has changed and Holmens is now part of MODO.

In Figure 7.2 the four stock prices are drawn. Notice in
particular the turn after time 100 where the values start to

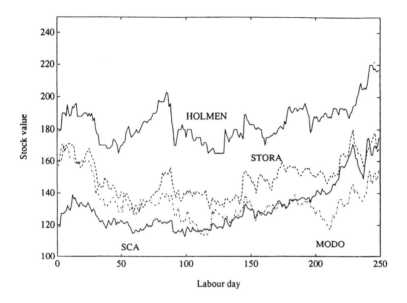

Figure 7.2 *Stock values for the companies STORA (1), SCA (2), MODO (3), and HOLMENS (4) during 1985.*

rise towards the end of the year. In Figure 7.3 we display as a general background the volumes of sold stocks (which were left out of the analysis), and in Figure 7.4 series 5–8 are rescaled and shown.

Suppose we want predictions of one of the series, SCA say, five time units ahead. Since we normally have five working days in a week, this typically means one week forecasts. In the following analysis the computations will be made anew for each new time point, so that the system always gives an updated prediction.

In our notation this means prediction of $X_2(t + \tau)$, $\tau = 5$, based on $\mathcal{F}_t$, where $\mathcal{F}_t$ is generated by the eight series up to time t. We will simplify this general setting and define our models as members of the class of multivariate, sparse autoregressive models (4.5) with linear trend and no seasonal effects. We write the forecasts of the model in the form

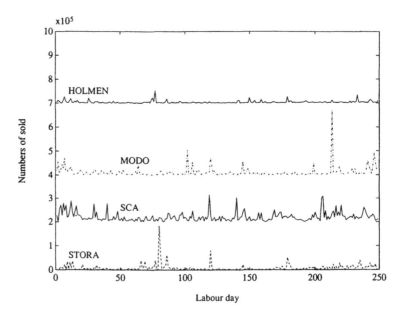

Figure 7.3 *Volumes of sold stocks for the four companies during 1985. Zero levels at 0, 2, 4, 7 times 10^5.*

$$\hat{X}_2(t + \tau) = \beta_0 + \beta_1(t + \tau) + \sum_{i=2}^{p-1} \beta_i X_{j_i}(t - t_i), \qquad (7.8)$$

where p is the model size, β_i are parameters to be estimated, j_i is the index of the ith selected predictor series, and t_i is the lag of the same predictor. We also restrict the number of available time points in order to reduce model selection errors. Somewhat arbitrarily the lags $t_i = 0, 1, 2, 5, 10$ are offered where lag 0 is the last available observation, and lag 10 is two weeks old. All eight series above can be used at these lags in the prediction.

All the linear combinations of this set of 42 allowed predictors, with no more than 15 predictors ($2 \leq p \leq 15$) and containing the trend $\beta_0 + \beta_1(t + \tau)$, are candidates in a model selection procedure.

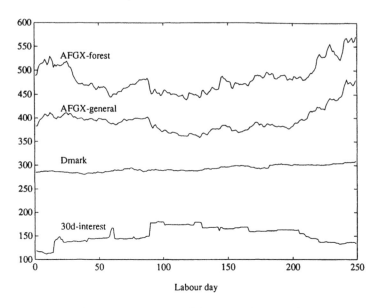

Figure 7.4 *Thirty days interest (5), general index (6), forest index (7), rescaled exchange rate Skr to Gmarks (8) during 1985.*

The parameter estimates for the prediction model use the data up to the present time t for each new forecast of $X_2(t + \tau)$. Consequently they are updated as time proceeds. However, since this model class is not a perfect one, and in particular the use of linear trends can be very poor over longer periods, we will allow adaptivity in our updated estimates of the models' parameters. The adaptivity is accomplished by a forgetting factor in the estimates of trends and covariances. The measures $C(\alpha)$ and CMF, see (4.6)–(4.7), are also added with the same forgetting factor to allow adaptivity in the model size decisions and the evaluations. At each time the model size minimizing C will be selected, and a model of that size is then determined by forward selection using the estimated covariances. A FORTRAN program performs the task very quickly on a work station and can also be run conveniently on a PC/AT for this and similar problems with reasonable sets of predictors.

One of the consequences of the adaptivity is that the model decisions and the estimates will never stabilize. This can be a useful property since it allows adaptation to changes in the financial system, but if the system were stable it would introduce some unnecessary noise in the analysis. The adaptation is of course not immediate, but has a delay determined by the forgetting factor. There is no strict definition of this delay, since all old data are weighted into the estimates. However, since the weights in our case were halved after about two months, the delay has the same order of magnitude.

In Figure 7.5 we show the variations with time of the model selection. Both the model sizes and the predictors are highly variable and give a clear impression of the randomness discussed in Chapters 2–4.

Another interesting comparison is given by the decision measure $C(\alpha)$ at different time points. The measure was defined in (4.6). Here the alternatives for α are the model sizes p. In Figure 7.6 we give the results at times $t = 100$, 150, 200, and 249. There are some difficult occasions in the late autumn of 1985 which cause large prediction errors and affect all the adaptive measures heavily at the end of the series. For example the CMF-measure becomes 7.366, 6.984, 6.944, 23.676 at times $t = 100$, 150, 200, 249.

We can see from Table 7.1 that $C(p)$ is very flat for $p \geq 4$ and has a minimum for $p = 10$. The AIC$'$ points at the largest model but has a local minimum at $p = 11$. This time it does not underestimate the prediction error variance as badly as it did with the shorter multivariate series in Chapter 4. The least squares points at the largest model almost by definition and indicates about half the uncertainty shown by CMF or $C(p)$.

Now to the interesting philosophical question. Will these forecasts improve on persistency, or is the stock system too close to a martingale system with a perfect market evaluation of present stock values all the time? Let $e_{t,\text{pers}} = X_2(t) - X_2(t - \tau)$ be the error of predicting unchanged values from time $t - \tau$ to time t. Write $\hat{X}_2(t|t - \tau)$ for the prediction of $X_2(t)$ by the model selected and estimated at

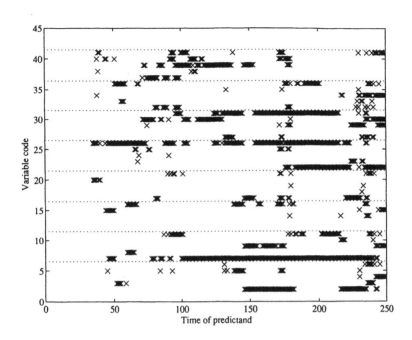

Figure 7.5 *Predictors selected for five times ahead SCA stock forecasts at different times. Predictor codes: 2–6 series 1 at lags 0, 1, 2, 5, 10; 7–11 series 2 at lags 0, 1, 2, 5, 10; ...; 37–41 series 8 at lags 0, 1, 2, 5, 10. Two trend variables are forced into all models and not shown.*

time $t - \tau$, $\tau = 5$, and let $e_{t,\mathrm{mod}} = X_2(t) - \hat{X}_2(t|t - \tau)$ be the corresponding prediction error. Take the difference of the squared prediction errors

$$d_t = e_{t,\mathrm{pers}}^2 - e_{t,\mathrm{mod}}^2.$$

The useful times are $35 \le t \le 249$ since 34 times are needed initially to produce all models. Checking the signs we find that d_t is negative 86 times and positive 129 times. If d_t are independent, this difference is significant at the 5% level. We can also compute $\bar{d} = 11.01$, and $s_d = 48.89$, and again an assumption about independent and identically distributed

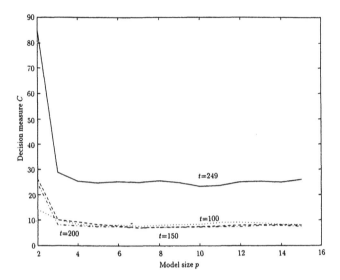

Figure 7.6 *The decision measure $C(\alpha)$ of SCA stock predictions at four times in the analysis, modified for adaptivity.*

variables would give a significant positive value at about the 1% level $(\bar{d}/(s_d/\sqrt{215}) = 3.3)$. However, we are in a time series situation and cannot expect independence for five days forecasts. Checking for correlations in the series d_t we find the first three estimated as $(r_1, r_2, r_3) = (0.53, 0.11, 0.05)$, and they are not uniformly small after lag three. Assuming stationarity, it is elementary to show that the variance of $\bar{d}$ now becomes

$$\mathrm{Var}(\bar{d}) = \frac{\sigma^2}{n}(\rho_0 + 2\rho_1(1 - \frac{1}{n}) + 2\rho_2(1 - \frac{2}{n}) + \ldots),$$

where ρ_i is the true correlation at lag i, and $n = 215$ is the number of observations. Using estimates r_i for ρ_i the variance more than doubles (the exact value is dependent on where the formula is truncated), and the significance is no longer clear.

Although we have some indications that it is possible to make better forecasts than just predicting no change, a one

Table 7.1 *Final variance estimates at time t = 249 for least squares, AIC', and the forward validation decision measure.*

p	LS	AIC'	$C(p)$
2	80.4	85.2	84.8
3	23.1	25.2	28.8
4	20.0	22.4	25.3
5	18.7	21.6	24.7
6	17.7	21.0	25.2
7	16.3	19.9	24.8
8	15.8	19.9	25.5
9	15.4	19.9	24.7
10	14.8	19.8	23.3
11	13.9	19.1	23.7
12	13.6	19.1	25.0
13	13.3	19.3	25.3
14	12.9	19.3	24.9
15	12.0	18.4	26.0

year record of data is not enough to prove this with significant results, at least not for the model class we have used. If better predictor series can be defined and the improvement compared to 'unchanged' therefore becomes larger this may be shown by the corresponding analysis on a data series of this size. We leave this as a challenge for the reader. Remember however that if many different attempts are made to find this golden result, the lesson from the earlier chapters is that the full model selection with the different approaches should be validated together, since we cannot rely on the best selected result itself. Of course, if we move such a test to new and independent data a simpler validation can be allowed.

7.3 Model selection and validation in meteorology

The atmosphere is a complex object. It obeys some general physical laws such as Newton's laws for the large scale transports of the air masses, and also some laws for radiation and other kinds of exchange of energy. At the ground, where most observations are taken, local effects will disturb the general phenomena, and higher up in the atmosphere the clouds and water exchange can largely not be predicted by simple physical equations. Meteorologists often compare their numerical weather prediction models with the now so-popular chaotic models. A minor change in the initial conditions may have dramatic effects on the model's forecasts a few days later. From a probabilist's point of view it is more natural to think of models with random noise added to some conceived differential equations.

In both cases the information about the future will be of a statistical nature, and when it comes to realistic problems, a more basic regression or time series approach is often the most reasonable way to useful results. One problem is that so many special mechanisms are combined in the same problem and can motivate a large set of predictors in order to describe them all. Often, it is not easy to tell which are important and which are not. The problem may therefore become one of model selection from a huge set of possible models. Fortunately the amount of data is large too. In Hjorth and Holmqvist (1981) the following analysis was made on data from Linköping and eight surrounding places during winter (Jan–Feb) 1971–78.

7.3.1 Covariance and linear prediction

Let $Y_{rn}(t)$ denote variable r at time t, $1 \leq t \leq T$, and year n, $1 \leq n \leq N$. In this case we have $T = 59$, and $N = 8$. Estimate all means and covariances as

$$\overline{Y}_r = \frac{1}{NT} \sum_{n=1}^{N} \sum_{t=1}^{T} Y_{rn}(t)$$

$$\widehat{R}_{rs}(\tau) = \frac{1}{NT} \sum_{n=1}^{N} \sum_{t=1}^{T-\tau} (Y_{rn}(t+\tau) - \overline{Y}_r)(Y_{sn}(t) - \overline{Y}_s)$$

for non negative τ, and let $\widehat{R}_{rs}(\tau) = \widehat{R}_{sr}(-\tau)$ when $\tau < 0$.

Let V be a set of indices corresponding to the variables of interest, and denote by $\mathbf{R} = \{\widehat{R}_{rs}(\tau); r, s \in V\}$ the set of estimated covariances. The covariances can be compared to the matrix $\mathbf{X}'\mathbf{X}$ in regression when the means are subtracted from all variables and the Y-variable is included among the X's. We can therefore define stepwise regression procedures just like we do in ordinary regression. Let

$$FS(p,\ Y_i(t+\tau)\ \mid\ (V,\lambda)\mid \mathbf{R}) \qquad (7.9)$$

denote p steps of a Forward Selection procedure (FS) for the prediction of Y_i τ time-steps ahead, and without stopping rules based on F-tests. Here (V,λ) stands for the set $\{Y_r(t-l), r \in V,\ 0 \leq l \leq \lambda\}$ of potential predictor variables, and $\mathbf{R}$ for the covariances used. The result is a prediction model

$$\widehat{Y}_{ip}(t+\tau) = \overline{Y}_i + a_{p1}(Y_{r_1}(t-l_1) - \overline{Y}_{r_1}) + \ldots$$
$$+ a_{pp}(Y_{r_p}(t-l_p) - \overline{Y}_{r_p}) \qquad (7.10)$$

and an estimated prediction-error variance given as

$$\widehat{\sigma}_{ip}^2(\tau) = \widehat{R}_{ii}(0) - \sum_{m=1}^{p} a_{pm}\widehat{R}_{ir_m}(l_m + \tau). \qquad (7.11)$$

7.3.2   Validation

Within each year the data are dependent, but observations from different years can be considered independent on the time scales we are interested in. (In a very long perspective, the observations are from the same climate period and can therefore be considered dependent. However, our models will not be used forever, and we are not troubled by such dependency as long as we use the model under approximately the same climate.) An obvious possibility is to use the independent years for a validation. We then take out one year as test set each time, and do all the above calculations on the remaining seven years. Let year k be the test set and denote by $\overline{Y}_{r;-k}$, $\widehat{R}_{rs;-k}(\tau)$, $\mathbf{R}_{-k}$ the mean, covariance, and set of covariances for the corresponding estimation set. We replace of course N by $N-1$ in the formulas when estimates are based on so many years. Compute the mean square prediction error on the test set for each model size p, and denote this by

$$
\mathrm{CV}_{-k}(p) = \frac{1}{T - \tau - \lambda} \sum_{t=\lambda+1}^{T-\tau} (\widehat{Y}_{ip;-k}(t + \tau) - Y_{ik}(t + \tau))^2.
$$

Then take the average over all possible test sets (years) to get

$$
\mathrm{CMV}(p) = \frac{1}{N} \sum_{k=1}^{N} CV_{-k}(p). \tag{7.12}
$$

The selected model size will be the p_0 such that $\mathrm{CMV}(p_0) = \min_{1 \le p \le q} \mathrm{CMV}(p)$ for some chosen upper limit q. Returning to the covariances $\mathbf{R}$ of the complete data, we compute the prediction model $\widehat{Y}_{ip_0}(t+\tau)$ as in (7.10) and use $\mathrm{CMV}(p_0)$ as our estimate of the prediction error variance. In the referred study, no attempt was made to let p_0 vary with the different test sets.

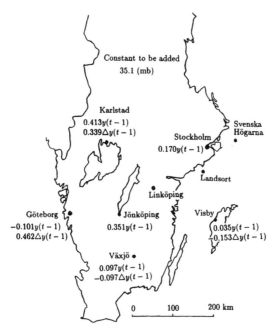

Figure 7.7 *Map of south Sweden showing the nine data positions and the 6 hr prediction model for air pressure in Linköping. Hjorth and Holmqvist (1981).*

7.3.3   Illustration of results

Our first result is on air pressure. Let P denote the indices for air pressure measurements in the nine places shown in Figure 7.7. The set

$$(P, 19) = \{Y_r(t - l),\; r \in P,\; 0 \le l \le 19\}$$

contains 180 predictor variables. For pressure forecasts six hours ahead the $\mathrm{CV}_{-k}(p)$ functions are given in Figure 7.8. (Six hours means very short forecasts in meteorology). The average $\mathrm{CMV}(p)$ is also indicated by the dotted curve. A very flat minimum at $p_0 = 9$ suggests this model size. For the complete data set this prediction model uses observations from the last two time points only. Defining 6 hours as our time unit, and predicting the pressure at time t with $t - 1$ as the last observation time we arrive at the model given in

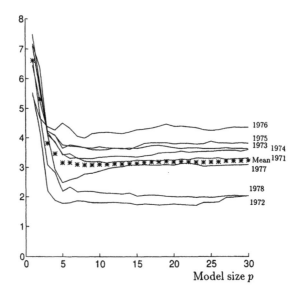

Figure 7.8 *Variances of 6 hr pressure predictions estimated by cross validation on each excluded year, and the mean* CMV(p) *as functions of model size p. Hjorth and Holmqvist (1981).*

Figure 7.7, where we use differences $\triangle y(t-1) = y(t-1) - y(t-2)$. The precision of linear forecasting appears to be very competitive on this short time scale and up to about 20 hours.

A corresponding analysis for temperature in Linköping had the extra complication that there were important seasonal components both for daily variations and for the seasons of the year. Also, much more complicated physics can motivate far more interesting predictors than for pressure. The following sets of variables were used in one study (time unit 6hr).

$(T_0, 49)$ temperature at Linköping with time lags
            up to 12 days, 50 variables;

$(T, 19)$ temperature at nine places with time lags
            up to 5 days, 180 variables;

$(TD, 19)$ the set above and four trigonometric terms
            with the day's and the year's periods;

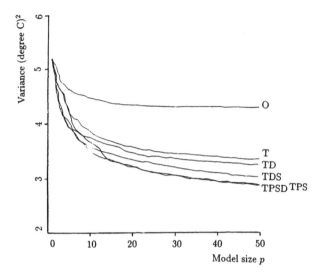

Figure 7.9 *Variances for 6 hr temperature predictions estimated by the FS routine for different variable sets. Hjorth and Holmqvist (1981).*

$(TDS, 19)$ the last set and six combined functions of gradients for temperature and pressure with time lags, 304 variables;

$(TPS, 19)$ the trigonometric variables replaced by lagged pressure variables, 480 variables;

$(TPSD, 19)$ the above set and the trigonometric terms, 484 variables.

In Figure 7.9 we show the classical least squares measures, and in Figure 7.10 we show the cross validation measures as function of the model size p. No doubt the validation gives quite different and valid information. The classical measure gives no reliable idea of which model size to use.

The present investigations were part of a mapping of a problem area. Many useful averages could be made before the analysis in order to bring down the number of predictors in the analysis. Typically we will eliminate some model selection and estimation noise that way. Here it is interesting to notice that rather large models were selected.

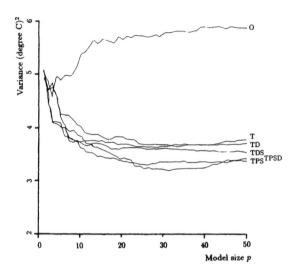

Figure 7.10 *Variances for 6 hr temperature predictions estimated by cross validation for different variable sets. Hjorth and Holmqvist (1981).*

7.3.4   Validating the need of data

Questions about the need of data for a specific analysis are among the most difficult to answer in statistical practice, especially when the situation is a bit complex. When we already have some data, we can instead ask if these data are sufficient, or if much more could be gained by more data. We may analyse this backwards by observing the effects when we reduce our data set, but it is not trivial to arrange this comparison in an informative way. I will give one example from a meteorological study. The same kind of illustration could be made on most analyses of the regression type, or more generally on analyses where a valid prediction and cross validation or bootstrap evaluation can be defined. For a sharp comparison it is very important to maintain some kind of symmetry so that all data are predicted even when smaller data sets are studied.

In the eight year data base for pressure prediction used earlier, we now subdivide our data into two groups of four years

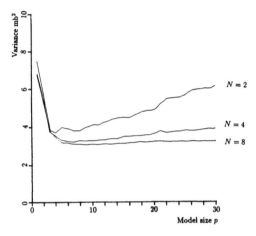

Figure 7.11 *Mean variances for 6 hr pressure predictions estimated by cross validation for N = 2, 4, and 8 years. Hjorth and Holmqvist (1981).*

each. We then make a full cross validation study in each group of four years. Averaging the cross validation measures for the two groups we get the results of predicting each year once with a data base of four years.

A full symmetry would require that, with four years, each year should be predicted by all sets of 3 other years. This is possible when we have in all eight years, but we get 280 different combinations of predictors and predictand. We were lazy enough to use the simpler approach in our study.

We can also divide the data base into four groups of two years each, and make a cross validation within each pair of years. Averaging these results we get an evaluation of data bases with only two years of observations.

The result is shown in Figure 7.11 and gives the impression that a further doubling of the data would not give much improvement of the prediction model. It also shows that the step from $N = 2$ to $N = 4$ was very important, and that the step from 4 to 8 has been worthwhile. A more general use of such pictures could give very good information to data collectors and modellers.

In an ordinary regression problem we may subdivide our data randomly or systematically, but it is a good idea to still predict all values equally often for all compared sizes of the data base.

This kind of study is also very handy in bootstrap problems, if we have independent and identically distributed units of data. We simply vary the size of the resamples and study the effects on the estimated uncertainties. In complex analyses this will not always obey the $1/\sqrt{n}$-law which is so typical for traditional parameter estimates.

7.3.5   Validation of rain probability forecasts

Predicting rain provides some interesting problems and we will describe some experiences of work with such forecasts. The goal was to produce predictions for a whole area and to display such predictions on maps. We will not enter into all aspects here, but we found it useful to put together data from several different places within the area, and estimate one model for all of them. Instead the predictors did account for geographical factors. This model could then be used also for places not represented in the data base. After the determination of a proper time resolution we have to face the fact that rain occurs in a minority of the cases. Six hours turned out to be the shortest useful interval in our data base, and about 25% of the intervals did show measurable precipitation in one of our studies. It is therefore natural to divide a rain model into two parts where part one describes the probability for rain and part two gives a conditional distribution for the amount given that it rains. We will discuss step one only.

Let $X_1, \ldots, X_m$ be predictors with information relevant to the probability of rain. Some of these variables represented predicted humidity, cloudiness, vertical wind, etc. in the global numerical model produced at the European centre ECMWF in Reading, and others represented more local and recent data from the meteorological observations around and 'transported' into our area with the wind predicted in the nu-

for the entire set of data. We use Q_2 as a measure of fit just like we use least squares as a distance measure regardless of whether any normal distribution and indepence assumption will give a likelihood argument for it. The logit model is fitted to data by maximizing numerically the measure Q_2. We can also arrange stepwise selection of variables and cross validation in the same way as we did in ordinary regression.

The above analysis is however far more complicated in a generalized linear model when the set of predictors becomes large. We may enter problems with the numerical solution of estimates, which we never do for a non-singular linear model. The following steps are therefore sometimes of value. Start with a critical examination to see if some predictors can be removed. Next, group the predictors into groups carrying related information and see if the groups can be represented by a few of the variables or some linear combinations. There are some standard tools for such reductions, such as principal components. These tools should be used with great care, but with a logical motivation behind them they can be useful. On large data bases or with complicated models one may sometimes use approximations and tricks which are not so clean, and seldom show up in the scientific literature. The following is a useful one. Make ordinary stepwise linear regression of y against $X_1, \ldots, X_m$ by the forward selection method and use cross validation to decide how many and which predictors to use. The procedure in Chapter 3 is applicable. Then take these predictors into the generalized linear model without further efforts to select among them. We assume that the set is reasonably small, otherwise other dirty tricks will be used such as replacing a set of not so important variables by a linear combination (perhaps suggested by the linear regression) or restricting the search of these variable's parameters in some other way. The final analysis is a pure (but formal) maximum likelihood exercise on cases treated as independent. It is formal since several simultaneous cases from different places are in fact dependent. This does not harm the parameter estimates as long as the set of time points is large enough to give high precision. The precision will be lower than with the same number of independent data, but higher than if data are thinned so that only independent cases are used. An implementation of the model was reported in Häggmark and Hjorth (1987).

The precision was never estimated in our study. It was considered 'more than enough' for the purpose. It can in fact not be evaluated by classical methods, but we now have the solutions. The best evaluation would certainly be to resample among the time points and select or exclude simultaneous observations at all places. The bootstrap is then the natural tool. Since a proper evaluation should involve also step one, where predictors were selected by forward selection and cross validation, this would be a large exercise. We did not find this worthwhile, and also resisted any attempt to lie about the parameters' exact precision by simpler methods.

7.4 Bootstrapping a cost function

Some cost functions are based on statistical information. It is then important to know how the uncertainties of the data will show up in the cost functions, and bootstrap is one tool for doing this. Usually one only seeks the minimum cost or, perhaps more often, the strategy minimizing the cost function. For sensitivity analyses and for compromises with other interests, the whole cost function will be needed. We will look at an optimal replacement problem. This is a situation where the cost can sometimes be a function of one single argument, the replacement age.

7.4.1 The replacement problem

A mechanical unit will be replaced at failure or at the age T, whichever comes first. Several such units have been working for some time and many of them have already failed, so we have some data on the life time distribution. There are also some costs related to the failures and replacements of these units.

Let

C_1 = cost for replacement without failure;
C_2 = cost for failure and replacement of unit.

In C_2 all secondary costs like damage on other units, delay
and loss of production are included. Typically $C_2 > C_1$
and sometimes by a large factor. We suppose C_1 and C_2
are known constants. If the costs are random we use the
expected costs. (When C_1 and C_2 are difficult to set, a
useful approach is to estimate a set of cost curves for some
ratios C_2/C_1 and take C_1 as cost unit.)

Let $F(x)$ be the life time distribution of the unit, and
introduce the *survival function* $R(x) = 1 - F(x)$ giving the
probability to survive age x. Replacing non-failed units at
age T gives an expected cost per unit

$$F(T)\,C_2 + (1 - F(T))\,C_1 = R(T)C_1 + (1 - R(T))C_2.$$

The expected working time for one unit, truncated at time
T, is given as

$$\mu(T) = \int_0^T x\,f(x)\,dx + T(1 - F(T))$$
$$= \int_0^T (1 - F(x))\,dx = \int_0^T R(x)\,dx,$$

where the last integrals are correct also for distributions
without a density. Renewal theory shows that the ratio of
these two expectations will give the average cost per time
unit in the long run (Ross 1989, Chapter 7). If we want to
minimize this long term average, the natural cost function is
given as

$$C(T) = \frac{R(T)\,C_1 + (1 - R(T))\,C_2}{\int_0^T R(x)\,dx} \qquad (7.14)$$

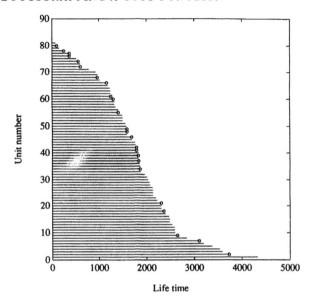

Figure 7.12 *Life time data of 81 units, with censoring illustrated by balls.*

7.4.2 Estimation

A typical set of data will consist of a mixture of times to failure and times to censoring. The censoring will in turn be a mixture of replacement ages and times without failure for units which have neither been replaced nor failed yet. For the correct estimation, it is important to remember all units which have not failed. They may contribute a lot to the survival probability, since they have in fact survived up to their present age. This may seem trivial in theory, but in practice it requires a very clear definition of the studied population of units. From such data we have to estimate the survival function $R(x) = 1 - F(x)$. This can be done by using parametric distributions and estimating the parameters, for example, by maximum likelihood. We can also use extensions of empirical distribution functions. Since the traditional bootstrap is defined in terms of empirical distribution functions, these are of special interest.

Let $\widehat{R}(x) = 1 - \widehat{F}(x)$ be an estimated survival function. The estimated cost curve will then be given by

$$\widehat{C}(T) = \frac{\widehat{R}(T)\, C_1 + (1 - \widehat{R}(T))\, C_2}{\int_0^T \widehat{R}(x)\, dx}. \qquad (7.15)$$

Data from 81 units are displayed in Figure 7.12. For each unit, a line is drawn as long as it has been working. A ball at the end of the line shows censoring. There are 24 censored cases. Different estimated survival functions are drawn in Figure 7.13. One of them is called the continuous product limit, CPL, and is a continuous version of an empirical survival function. The other two are the standard discontinuous empirical survival function (1− the empirical distribution function) introduced for censored data by Kaplan-Meyer (1958) (see also Lawless 1982), and the two-parameter Weibull distribution estimated by maximum likelihood. See the Appendix in Section 7.4.4 and also Section 6.6.1 for details about the estimates.

Let the planned replacement cost be our cost unit. The average cost for failure was estimated to be 15 times as high. We then set $C_1 = 1$ and $C_2 = 15$. The estimated cost function is drawn (solid) in Figure 7.14 for the CPL-estimated survival function. It is there surrounded by a sample of other cost functions described in the next section.

7.4.3 Bootstrap resampling

Three different resampling methods will be given. The choice of procedure will depend on how rich our data are, and on our ideas about how they were generated, and on other background information like knowing that the distribution is continuous and smooth. Of particular interest for the resampling is whether we are willing to model the censoring in some way.

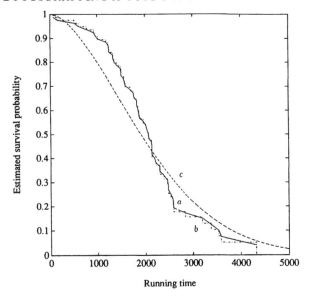

Figure 7.13 *Survival function estimated by   a: continuous product limit,   b: Kaplan-Meyer's estimator,   c: maximum likelihood Weibull distribution.*

Method 1

The simplest bootstrap method, and often good enough when we have a relatively rich set of observations, is to re-sample at random from all data. For the 81 cases shown in Figure 7.12, this is equivalent to making 81 independent draws of an integer between 1 and 81, and using the corresponding observation as our bootstrap observation. Each case will be a failure or censored just as in the original data. The number of censored data values will then vary at random (as a Binomial variable) during the resampling. For the bootstrap sample we compute estimates $\widehat{R}^*(x)$ and $\widehat{C}^*(T)$ in exactly the same manner as for the original data.

A sample of bootstrap cost functions generated by this method is drawn in Figure 7.14. In order to keep the picture readable, only ten resamples are displayed. The bootstrap sample indicates some interesting facts. The cost curve is rather uncertain. At the minimum, the highest cost mini-

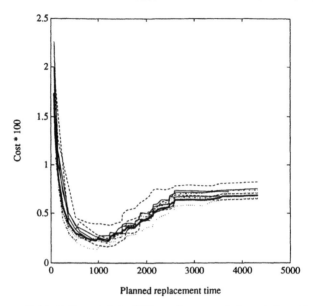

Figure 7.14 *Estimated (solid) and resampled cost functions using the* CPL *estimate.*

mum is almost three times as large as the lowest bootstrap generated cost minimum. On the other hand, the general shape of the cost curve appears to be rather well estimated. The different bootstrap generated minima will for example indicate strategies whose costs are fairly close to the optimal cost for the solid cost function. This means that we will probably estimate close to optimal strategies, although we are very uncertain of their true cost. This nice property was found in some simulation studies before the bootstrap was published. Part of the results were later published in Borgefors and Hjorth (1981). Today bootstrap would be an excellent tool in investigations of such an empirical nature.

Method 2
The second resampling method for this problem is conditional on the number of failures. When failures are few, the randomness caused by varying the number of failures in the resample may be misleading. In practice we have

perhaps collected data until a reasonable number of failures were observed. Resamples with much smaller sets of failures do not reflect this. We may then resample at random from all data until n failures occur instead of resampling a fixed number of observations as in method 1. The number of censoring times, m^*, is now random and the total resample size $n + m^*$ gets a negative binomial distribution, $NB(n,p)$, with probabilities $p_k = \binom{k-1}{n-1} p^n (1-p)^{k-n}$, $k = n, n+1, \ldots$ and $p = n/(n+m)$. The same resampling can therefore also be achieved by first sampling a number k from the $NB(n,p)$-distribution and then resampling n failure times from the observed failures and $m^* = k - n$ censoring data from the original censored observations.

This method 2 supposes that the data collection stops just after the nth failure. Otherwise, more detailed information about the stopping rules for the data collection will be needed.

Data: $t_1, \ldots, t_n$ n failure times
       $u_1, \ldots, u_m$ m censoring times

Data*: $t_1^*, \ldots, t_n^*$ drawn independently from $t_1, \ldots, t_n$
       $u_1^*, \ldots, u_m^*$ drawn independently from $u_1, \ldots, u_m$

Method 3
Sometimes we may consider a distribution for the censoring times. One concrete reason may be that the failures on some other unit will cause the censoring. We can then estimate two survival functions (or distribution functions) $R_f(x)$ and $R_c(x)$ for the failures and the censoring times respectively. Parametric or non-parametric estimates are possible.

Let $\widehat{R}_f(x)$ and $\widehat{R}_c(x)$ be the continuous version of the empirical distribution. The following bootstrap generation can then be applied. Let U_i denote random numbers which are uniform in $(0,1)$. Give two such numbers U_1 and U_2. Compute

$$X = \begin{cases} \widehat{R}_f^{-1}(U_1) & \text{if } U_1 > \widehat{R}_f(t_n), \\ t_n & \text{otherwise}; \end{cases}$$

$$Y = \begin{cases} \widehat{R}_c^{-1}(U_2) & \text{if } U_2 > \widehat{R}_c(u_m), \\ u_m & \text{otherwise}. \end{cases}$$

If $X \leq Y$, $X < t_n$, $Y < u_m$: resample a failure at X;

if $Y < X$, $X < t_n$, $Y < u_m$: resample a censoring at Y;

otherwise resample a censoring at t_n. We have here assumed that data are ordered as above so $t_1 < t_2 < \ldots < t_n$ and $u_1 < \ldots < u_m$.

A parametric version of the same bootstrap method was discussed in Section 6.6.1, where we assumed Weibull distributions for both the failures and the censorings.

7.4.4 Appendix: A continuous empirical survival function

This estimate is a small modification of the Kaplan–Meier (KM) estimator, Kaplan and Meier (1958), and we will present the old estimator as a background. The KM estimator is a well-known empirical survival function for censored data. It uses products of factors and is also called a product limit estimator.

Let $t_1 < t_2 < \ldots < t_n$ denote the ordered set of failure times. Our time concept is usually the age in working hours of the units. Let $R(x) = 1 - F(x)$ be the probability that a unit survives time x. Start with $\widehat{R}(0) = 1$. Denote by $n(t)$ the number of units still working at their age t. Let t^- and t^+ denote just before and just after time t. At any failure time t_i the KM estimator sets

$$\widehat{R}_{KM}(t_i^+) = \widehat{R}_{KM}(t_i^-) \frac{n(t_i^-) - 1}{n(t_i^-)} \qquad (7.16)$$

If more than one unit fails at the same time t_i, the figure 1 will be replaced by the number of failures (Lawless 1982). Between failure times this estimator is constant. The

Kaplan–Meier estimator has reasonable asymptotic properties under certain assumptions about the censoring, but one
can criticize it in finite and censored cases for not using information about when the censoring takes place between failures. The estimator is also typically biased in finite censored
samples. Suppose our model is a continuous distribution and
that we want a continuous estimate. Then a linear interpolation between failure times will do. We prefer however to
interpolate between unique and if possible unbiased values.
Without censoring, the expected value of $R(t_i)$ at failure
time i from a sample of n units will be $1 - \frac{i}{n+1}$. This follows since the sequence $R(t_n), \ldots, R(t_1)$ is distributed as the
ordered sample from n independent variables uniformly distributed on $(0,1)$. (This in turn depends on the fact that
for any variable X with continuous distribution function
F the random variable $F(X)$ is uniform in $(0,1)$ and so is
$R(X) = 1 - F(X)$.) Now

$$\mathbf{E}[R(t_i)] = 1 - \frac{i}{n+1} = (1 - \frac{1}{n+1})(1 - \frac{1}{n}) \cdots (1 - \frac{1}{n-i+2}),$$

where the denominator in each factor is one plus the number
of units still surviving before respective failure time $t_1, \ldots, t_i$.
This number is also the average number of units alive in the
interval between the two last failures. If we stick to this
average with censored data, we get a heuristic but sound
use of the information about when the censoring takes place
between failures. Let $t < u$ and introduce the average

$$M(t,u) = \frac{1}{u-t} \int_t^u n(x)\, dx = \frac{\sum \text{time in } (t,u) \text{ for unit } i}{u-t}.$$

Set $t_0 = 0$, $\widehat{R}(0) = 1$ and

Table 7.2 *Pairs of classified mea-*
sures from human backbones;
1083 data values.

y	x			
	0	1	2	3
0	821	111	54	10
1	10	13	18	9
2	2	4	6	15
3	4	2	2	2

$$\widehat{R}(t_i) = \widehat{R}(t_{i-1}) \frac{M(t_{i-1}, t_i)}{M(t_{i-1}, t_i) + 1}, \quad i = 1, 2, \ldots \quad (7.17)$$

at the failure times. Complete the estimate up to the last
failure time by linear interpolation between neighbouring t_i.
Call this a continuous product limit estimator (CPL estimator).

7.5 A backbone bootstrap

In a medical study, the relation between two different measures x and y was of interest. Both measures were taken on
the same vertebrae in human backbones, and were classified
into four categories each. The measure x was relatively easy
to get, and the measure y was more difficult but also more
established. The first modest aim was to show a dependency
between the measures. The data were given to the statisti-
cian on the form shown in Table 7.2, and consisted of 1083
pairs of values.

If all the 1083 data values were independent, a simple chi-
square test would reject the hypothesis that the two measures
are independent. However, it turned out that the data were
taken from only 94 individuals, so that several pairs of mea-

surements were from the same person, and also more than one from the same vertebra. This could of course introduce a strong dependency between data from the same person. One might consider reducing the data into one pair of measures for each person but this would result in an unacceptable loss of information.

The situation can be analysed by the bootstrap method in the following way. Let $D_1, \ldots, D_n$, $n = 94$, denote the sets of data from the persons $1, \ldots, n$. Also denote by DS= $\{D_1, \ldots, D_n\}$ the full set of data. Removing the identification of the individuals we can represent the data set as DS= $\{(x_i, y_i), i = 1, 2, \ldots, 1083\}$. This is the representation illustrated in Table 7.2.

Let # denote 'the number of', and define (with reference to the table)

$$\hat{p}_{ij}(\text{DS}) = \frac{\text{\# data in row } i \text{ column } j}{\text{\# data},}$$

$$\hat{p}_{i.}(\text{DS}) = \sum_j \hat{p}_{ij},$$

$$\hat{p}_{.j}(\text{DS}) = \sum_i \hat{p}_{ij}.$$

Compute $\hat{r}_{ij}(\text{DS}) = \hat{p}_{ij}(\text{DS}) - \hat{p}_{i.}(\text{DS})\hat{p}_{.j}(\text{DS})$ as a measure of the dependency. As an example, we have

$$\hat{p}_{12}(\text{DS}) = \frac{18}{1083},$$

$$\hat{p}_{1.}(\text{DS}) = \frac{50}{1083},$$

$$\hat{p}_{.2}(\text{DS}) = \frac{80}{1083}.$$

and

$$\hat{r}_{12}(\text{DS}) = 0.0132.$$

If $\hat{r}_{ij}(\text{DS})$ is significantly different from zero for some i and j, we have proved dependency.

7.5.1 Resampling procedure

In the resampling we will use the natural assumption that data from different individuals are independent, and that the data sets D_i are statistically equivalent. The amount of data from each person is then regarded as part of the random system. Draw a bootstrap sample of n individuals, with replacement as always, and let

$$\text{DS}^* = \{D_1^*, \ldots, D_n^*\} = \{(x_i^*, y_i^*), i = 1, \ldots, I^*\}$$

denote the data from these individuals. Compute

$$\hat{r}_{ij}^* = \hat{r}_{ij}(\text{DS}^*)$$

and study the variability of this measure. For the special case $i = 1$, $j = 2$, we get the result shown in Table 7.3.

With high probability, the bootstrap values are within ± 0.010 from the $\bar{r}_{12} = \hat{r}_{12}(\text{DS}) = 0.0132$. We therefore conclude that $\hat{r}_{12}(\text{DS})$ is significantly positive, and dependency follows.

Notice that the whole analysis depends on linking the data to the individuals in order to find independent data units. It can therefore not be reconstructed from the data table in the beginning of this section. This is a very common situation in medical statistical consultations where the medical researcher does some preliminary work on the data before he wants a statistical verification of his findings. The danger of using some ready made computer program for contingency tables without noticing the true nature of the data is obvious. When the realistic analysis is complicated, the computer intensive method becomes very attractive.

The material in this section is based on work by Nordgaard (1988) together with the author.

Table 7.3 *Grouped results from 1000 bootstrap simulations of $\hat{r}_{12}^*$.*

Midpoint	Counts
0.004	2
0.006	14
0.008	63
0.010	138
0.012	255
0.014	239
0.016	182
0.018	77
0.020	21
0.022	7
0.024	1
0.026	0
0.028	1

The list of applications can be continued by several more cases, but this book has to come to an end. The best continuation is up to the reader who can apply the ideas on her or his own fascinating problems, or come up with some new theory and methodology in the same spirit. Thank you for your company and do not hesitate to send your reports; they may inspire some new section in a possible second edition.

References

Akaike, H. (1969) Fitting autoregressive models for prediction. *Ann. Inst. Statist. Math.*, **21**, 243–247.

Akaike, H. (1973) Information theory and an extension of the maximum principle. *Proc. 2nd Int. Symp. Information Theory*, Akademia Kiado, Budapest, 267–281.

Akaike, H. (1977) On entropy maximization principle. *Applications of statistics*, ed. Krishnaiah, P. R., North-Holland Publishing Company, 27–41.

Allen, D.M. (1971) Mean square error of prediction as a criterion for selecting variables. *Technometrics,* **13**, 469–475.

Babu, G.J. and Singh, K. (1983) Inference on means using the bootstrap. *Ann. Statist.* **11**, 999–1003.

Babu, G.J. and Singh, K. (1985) Edgeworth expansion for sampling without replacement from finite populations. *J. Multivar. Anal.* **17**, 261–278.

Barndorff-Nielsen, O.E. and Cox, D.R. (1989) *Asymptotic Techniques for Use in Statistics.* Chapman and Hall, London.

Basawa, I.V., Mallik, A.K., McCormick, W.P. and Taylor, R.L. (1989) Bootstrapping explosive autoregressive processes. *Ann. Statist.*, **17**, 1479–1486.

Beran, R. (1987) Prepivoting to reduce level error of confidence sets. *Biometrika*, **74**, 457–468.

Beran, R. (1988) Balanced simultaneous confidence sets. *J. Amer. Statist. Assoc.*, **83**, 679–697.

Beran, R. (1990) Refining bootstrap simultaneous confidence sets. *J. Amer. Statist. Assoc.*, **85**, 417–426.

Berk, K. (1978) Comparing subset regression procedures. *Technometrics*, **20**, 1–6.

Bhattacharya, R.N. and Ghosh, J.K. (1978) On the validity of the formal Edgeworth expansion. *Ann. Statist.*, **6**, 434–451.

Bickel, P.J. and Freedman, D.A. (1981) Some asymptotic theory for the bootstrap. *Ann. Statist.,* **9**, 1196–1217.

Billingsley, P. (1968) *Convergence of Probability Measures,* Wiley, New York.

Borgefors, G. and Hjorth, U. (1981) Comparison of parametric models for estimating maintenance times from small samples. *IEEE Trans. Reliability,* **R-30**, 375–380.

Bose, A. (1988) Edgeworth correction by bootstrap in autoregressions. *Ann. Statist.,* **16**, 1709–1722.

Box, G.E.P. and Jenkins, G.M. (1970) *Time Series Analysis, Forecasting and Control.* Holden Day, San Francisco.

Breiman, L. (1992) The little bootstrap and other methods for dimensionality selection in regression: X-fixed prediction error. *J. Amer. Statist. Assoc.,* **87**, 738–754.

Breiman, L. and Spector, P. (1992) Submodel selection and evaluation in regression. The X-random case. *Int. Statist. Review,* **60**, 291–319.

Chatfield, C. (1980) *The Analysis of Time Series.* Chapman and Hall, London.

Chung, K.L. (1974) *A Course in Probability Theory.* Academic Press, New York.

Cline, D. (1988) Admissible kernel estimators of a multivariate density. *Ann. Statist,* **16**, 1421–1427.

Copas, J.B. (1983) Regression, prediction and shrinkage. *J. R. Statist. Soc. B.,* **45**, 311–335.

Cox, D.R. and Hinkley, D.V. (1974) *Theoretical Statistics.* Chapman and Hall, London.

Cressie, N.A.C. (1991) *Statistics for Spatial Data.* Wiley, New York.

Csörgö, S. and Mason, D.M. (1989) Bootstrapping empirical functions. *Ann. Statist.,* **4**, 1447–1471.

Diaconis, P. and Efron, B. (1983) Computer-intensive methods in statistics. *Scientific American,* **248**, 96–108.

DiCiccio, T.J. and Romano, J.P. (1988) A review of bootstrap confidence intervals. *J. R. Statist. Soc. B,* **50**, 338–354.

DiCiccio, T.J. and Tibshirani, R. (1987) Bootstrap confidence intervals and bootstrap approximations. *J. Amer. Statist. Assoc.,* **82**, 163–170.

Dobson A.J. (1990) *An Introduction to Generalized Linear Models.* Chapman and Hall, London.

Draper, N.R. and Smith, H. (1981) *Applied Regression Analysis*, second edition. Wiley, New York.

Efron, B. (1979) Bootstrap methods: Another look at the jackknife. *Ann. Statist.*, **7**, 1–26.

Efron, B. (1982) *The Jackknife, the Bootstrap, and Other Resampling Plans*. Regional Conference Series in Applied Mathematics, **38**. SIAM, Philadelphia.

Efron, B. (1985) Bootstrap confidence intervals for a class of parametric problems. *Biometrika*, **72**, 45–58.

Efron, B. (1987) Better bootstrap confidence intervals and bootstrap approximations. *J. Amer. Statist. Assoc.*, **82**, 171–185.

Efron, B. (1988) Computer-intensive methods in statistical regression. *SIAM Review*, **30**, 421–449.

Falk, M. and Kauffmann, E. (1991) Coverage probabilities of bootstrap-confidence intervals for quantiles. *Ann. Statist.*, **19**, 485–495.

Faraway, J.J. and Jhun, M. (1990) Bootstrap choice of bandwidth for density estimates. *J. Amer. Statist. Assoc.*, **85**, 1119–1122.

Feller, W. (1971) *An introduction to probability theory and its applications*. Wiley, New York.

Franke, J. and Härdle, W. (1992) (report 1987) On bootstrapping kernel spectral estimates. *Ann. Statist.*, **20**, 121–145.

Freedman, D.A. (1981) Bootstrapping regression models. *Ann. Statist.*, **9**, 1218–1228.

Freedman, D. (1984) On bootstrapping two-stage least-squares estimates in stationary linear models. *Ann. Statist.*, **12**, 827–842.

Häggmark, L. and Hjorth, U. (1987) Yttäckande sannolikhetsprognoser av nederbörd (Surface covering probability forecasts of precipitation). SMHI Promis Notes, **16**, Swedish Meteorological and Hydrological Institute, Norrköping.

Hald, A. (1952) *Statistical theory with engineering applications*. Wiley, New York.

Hall, P. (1986) On the bootstrap and confidence intervals. *Ann. Statist.*, **14**, 1431–1452.

Hall, P. (1988) Theoretical comparison of bootstrap confidence intervals. *Ann. Statist.*, **16**, 927–953.

Hall, P. (1992) *The Bootstrap and Edgeworth Expansion*. Springer, New York.

Hannan, E.J. and Quinn, B.G. (1979) The determination of the order of an autoregression. *J. R. Statist. Soc. B*, **41**, 190–195.

Harvey, A.C. (1989) *Forecasting, Structural Time Series Models and the Kalman Filter.* Cambridge University Press.

Hinkley, D. (1988) Bootstrap methods. *J. R. Statist. Soc B,* **50,** 321–337.

Hinkley, D. and Wei, B.-C. (1984) Improvements of jackknife confidence limit methods. *Biometrika,* **71,** 331–339.

Hinkley, D.V., Reid, N. and Snell, E.J. (1991) *Statistical Theory and Modelling,* Chapman and Hall, London.

Hjorth, U. (1982) Model selection and forward validation. *Scand. J. Statist.,* **9,** 95–105.

Hjorth, U. (1987) *Stokastiska processer, korrelations- och spektralteori.* Studentlitteratur, Lund.

Hjorth, U. (1989) On model selection in the computer age. *J. Statist. Plann. Inference,* **23,** 101–115.

Hjorth, U. and Holmqvist, L. (1981) On model selection based on validation with applications to pressure and temperature prognosis. *Appl. Statist.,* **30,** 264–274.

Hocking, R.R. (1976) The analysis and selection of variables in linear regression. *Biometrics,* **32,** 1–49.

Holm, S. (1990) Abstract bootstrap confidence intervals in linear models. *Scand. J. Statist.,* **20,** 157–170

Huber, P.J. (1964) Robust estimation of a location parameter. *Ann. Math. Statist.,* **35,** 73–101.

Hurvich, C.M. and Zeger, S.L. (1987) Frequency domain bootstrap methods for time series. Faculty of Business Administration, Working Paper Series 87–115, New York University.

Izenman, A.J. (1991) Recent developments in nonparametric density estimation. *J. Amer. Statist. Assoc.,* **86,** 205–224.

James, G.S. and Stein, C. (1961) Estimation with quadratic loss. *Proc. 4th Berkeley Symp.,* **1,** 361–379.

Jenkins, G.M. and Watts, D.G. (1968) *Spectral Analysis and its Applications.* Holden Day, San Francisco.

Junghard, O. (1990) Linear shrinkage in traffic accident models and their estimation by cross validation and bootstrap methods. *Linköping Studies in Science and Technology. Thesis No.* **205,** Linköping.

Kaplan, E.L. and Meier, P. (1958) Nonparametric estimation from incomplete observations. *J. Am. Statist, Ass.* **53,** 457–481.

Kendall, M. (1976) *Time series.* Griffin, London.

Kullback, S. (1968) *Information Theory and Statistics.* New York, Dover.

Künsch, H.R. (1989) The jackknife and the bootstrap for general stationary observations. *Ann. Statist.*, **17**, 1217–1241.

Lawless, J.F. (1982) *Statistical Models and Methods for Lifetime Data.* Wiley, New York.

Lehmann, E.L. (1983) *Theory of Point Estimation.* New York, Wiley.

Loh, W.-Y. (1987) Calibrating confidence coefficients. *J. Amer. Statist. Assoc.* **82**, 155–162.

Loh, W.-Y. (1988) Discussion of P. Hall's paper. *Ann. Statist.* **16**, 972–976.

Martin, M.A. (1990) On bootstrap iteration for coverage correction in confidence intervals. *J. Amer. Statist. Assoc.*, **85**, 1105–1118.

McCullagh, P. and Nelder, J. (1989) *Generalized linear models*, 2nd edn, Chapman and Hall, London.

Miller, A.J. (1990) *Subset Selection in Regression.* Chapman and Hall, London.

Miller, R.G. (1974) The jackknife: a review. *Biometrika*, **61**, 1–15.

Moulton, L.H. and Zeger, S.L. (1991) Bootstrapping generalized linear models. *Comp. Statist. Data Analysis*, **11**, 53–63.

Navidi, W. (1989) Edgeworth expansions for bootstrapping regression models. *Ann. Statist.*, **17**, 1472–1478.

Nordgaard, A. (1988) Statistisk jämförelse av två sjukdomsbilder i ryggraden – en bootstrapstudie (Statistical comparison of two pathological pictures in the backbone – a bootstrap study). Technical report in Swedish, Dept. of Math., Linköping University.

Nordgaard, A. (1990) On the resampling of a stochastic process using a bootstrap approach. *Linköping Studies in Science and Technology. Thesis No. 223*, Linköping.

Nordgaard, A. (1992) Resampling stochastic processes using a bootstrap approach. In Bootstrapping and Related Techniques, *Lecture Notes in Economics and Mathematical Systems*, **376**, Springer, 181–185.

Pierce, D.A. and Schafer, D.W. (1986) Residuals in generalized linear models. *J. Amer. Statist. Assoc.*, **81**, 977–986.

Rao, C.R. (1973) *Linear Statistical Inference and Its Applications*, Second edn, Wiley, New York.

Rencher, A. and Pun, F.C. (1980) Inflation of R^2 in best subset regression. *Technometrics*, **22**, 49–53.

Rissanen, J. (1978) Modeling by shortest data description. *Automatica*, **14**, 465–471.

Ross, S.M. (1989) *Introduction to Probability Models*, 4th edn, Academic Press, London.

Rubin, D.B. (1981) The Bayesian bootstrap. *Ann Statist.*, 9, 130–134.

Schwarz, G. (1978) Estimating the dimension of a model. *Ann. Statist.*, 6, 461–464.

Silverman, B.W. (1986) *Density Estimation for Statistics and Data Analysis*. Chapman and Hall, New York.

Singh, K. (1981) On the asymptotic accuracy of Efron's bootstrap. *Ann. Statist.*, 9, 1187–1195.

Stein, C. (1956) Inadmissibility of the usual estimator for the mean of a multivariate normal distribution. *Proc. 3rd Berkeley Symp.*, 1, 197–206.

Stone, M. (1974) Cross-validatory choice and assessment of statistical predictions. *J. Roy. Statist. Soc B*, 36, 111–133.

Stone, M. (1977) An asymptotic equivalence of choice of models by cross-validation and Akaike's criterion. *J. R. Statist. Soc. B*, 39, 44–47.

Stone, M. and Brooks, R.J. (1990) Continuum regression: Cross-validated sequentially constructed prediction embracing ordinary least squares, partial least squares and principal components regression. *J. R. Statist. Soc B*, 52, 237–269.

Wand, M.P., Marron, J.S., and Ruppert, D. (1991) Transformations in density estimation. *J. Amer. Statist. Assoc.*, 86, 343–353, with discussion 353–361.

Wu, C.F.J. (1986) Jackknife, bootstrap and other resampling methods in regression analysis. *Ann. Statist.*, 14, 1261–1295.

Index

Milton Keynes UK
Ingram Content Group UK Ltd.
UKHW031533071024
449327UK00005B/69

9 780412 491603